WITHDRAWN

Essays in Political Philosophy

Essays in Political Philosophy

R. G. Collingwood

Edited with an Introduction
by David Boucher

CLARENDON PRESS · OXFORD

1989

Oxford University Press, Walton Street, Oxford OX2 6DP

Oxford New York Toronto
Delhi Bombay Calcutta Madras Karachi
Petaling Jaya Singapore Hong Kong Tokyo
Nairobi Dar es Salaam Cape Town
Melbourne Auckland
and associated companies in
Berlin Ibadan

Oxford is a trade mark of Oxford University Press

Published in the United States
by Oxford University Press, New York

British Library Cataloguing in Publication Data
Essays in political philosophy.
1. Politics
I. Title   II. Collingwood, R. G. (Robin George)
320
ISBN 0–19–824823–7

Library of Congress Cataloging in Publication Data
Essays in political philosophy/edited with an
introduction by David Boucher.
Includes index.
1. Political science.   2. World politics—1900–1945.
I. Collingwood, R. G. (Robin George).   II. Title.
JC257.C63 1989   320'.01—dc20   89–34178
ISBN 0–19–824823–7

Set by CentraCet, Cambridge
Printed in Great Britain by
Courier International Ltd,
Tiptree, Essex

To
Emma and Lucy

Acknowledgements

IT is with pleasure and gratitude that I acknowledge my many and various debts to all those who, whether consciously or unwittingly, have contributed something to the final appearance of this collection of essays. In particular, I wish to thank Rex Martin for his comments upon a different project, which nevertheless had a bearing on my conception of this. I owe a special debt to Peter Nicholson, who is more generous than anyone I know with his time in offering careful and considered comments, for making some suggestions about what ought to be included in the volume. My friends, Tariq and Glyn Modood, with their little daughter Ghiz, have always made me welcome in Oxford on my many visits. I cannot thank them enough for their hospitality, and gratefully acknowledge the benefit I have gained from my many discussions with Tariq about Collingwood's ideas. Thanks are due also to James Connelly for his help and encouragement, and to Mrs W. G. Morton for allowing items from the Knox papers, University of St. Andrews to be included in this collection of Collingwood's political writings. Robert N. Smart, Keeper of the Manuscripts, at St Andrews, kindly checked the copies against the originals.

La Trobe University School of Social Sciences Research Fund Committee, and the Research School of Social Sciences, Australian National University generously supported my trips to Oxford. La Trobe University Library gave me very real support in tracking down the most obscure published references and obtained copies for my perusal. Colin Harris and his staff, in particular Lindsay and Oonagh, in the Modern Western Manuscripts reading room of the Bodleian Library, Oxford University, deserve the highest praise for all the help they have given me. The many publishers who hold the copyright of the published items responded remarkably quickly to my requests for permission to reprint and without exception answered in the affirmative. In this respect I would like to thank Chicago University Press, The Macmillan Press, Cambridge University Press, Oxford University Press, and the Editor of the Aristotelian Society. For permission to publish Sir Henry Jones's reader's report of *Truth and Contradiction* I am indebted to Mrs Jean Hunt, The Macmillan Press, and Chadwyck-Healey Ltd. who have produced an edition of the Macmillan archives on microfilm. The British Library owns the MS and has graciously consented to its publication.

Liz Byrne, Mary Zaccari, Marilu Espacio, and Elisa Krahe are responsible for typing the volume. Their skills and proficiency eased my burden considerably, and it is with pleasure that I express my appreciation.

This volume would not have been possible, of course, without the consent of Mrs Teresa Smith, Collingwood's daughter. The decision was a difficult one for her to make, and she is to be applauded for allowing the previously unpublished material to be included here. Her generous permission allows, for the first time since Malcolm Knox exercised his editorial judgement, the publication of substantive philosophical discussions from the manuscripts. I cannot take full credit for persuading Mrs Smith of the merits of such a project, because Angela Blackburn of Oxford University Press played a considerable role in assuring Mrs Smith that the volume as a whole deserved to be published. I have been extremely fortunate in being entrusted to the editorial care of Angela Blackburn who has provided much needed encouragement throughout every stage of this project.

Finally, and once again, I thank my wife Clare for her unfailing support, and my daughters Emma and Lucy, to whom I dedicate this book, for providing the necessary distractions from the self-indulgence of academic work.

D.B.
Australian National University

Contents

I

Introduction

I

It is no longer true to say, as was once commonplace, that Collingwood is an unduly neglected thinker. This great polymath of the twentieth century produced such a diversity of literature that the aspirant postgraduate student can invariably find some new line of inquiry to pursue: a tendency considerably enhanced by the release of Collingwood's unpublished papers to the Bodleian Library. To write about Collingwood and his thought is, of course, in direct contravention of his exhortation not to waste time writing about him, but to write about the subject-matter and problems which he addressed.[1] Such is Collingwood's reputation and esteem that he has become subject-matter in his own right, and his theories the problems which we address. Without wishing to be flippant, or to suggest the mutual exclusiveness of categories, we have what may be described (to borrow a distinction from the natural sciences) as pure and applied Collingwood studies: the former concerned to understand, in all its complexity, what Collingwood means, and how he developed his conclusions over time; while the latter interrogates Collingwood's conclusions, modifies and develops them, and applies them to concrete philosophical problems.[2]

While not wanting to cast aspersions on analytic philosophy (indeed its practitioners are responsible for originally stimulating Collingwood studies), it is unfortunate that the analytic turn of mind has tended to dominate: unfortunate in that the principles in terms of which Collingwood is analysed and criticized are at variance with

[1] R. G. Collingwood, *An Autobiography* (London, Oxford University Press, 1970: first published 1939), 118.
[2] W. J. van der Dussen, *History as a Science: The Philosophy of R. G. Collingwood* (The Hague, Martinus Nijhoff, 1981), and R. Martin, *Historical Explanation: Re-enactment and Practical Inference* (Ithaca and London, Cornell University Press, 1977) are representative examples of pure and applied Collingwood studies respectively. The concerns are complementary, and neither is illegitimate, nor takes priority over the other.

his own principles. For example Collingwood is treated as a philosopher who has put forward various propositions about historical knowledge which invite interrogation. In other words he is treated as a philosopher who subscribed to propositional logic and is thus subjected mercilessly to the dissolving gaze of philosophical criticism which reveals for its practitioners that Collingwood was confused and contradictory.[3] The fact that he considered one of his most fundamental contributions to philosophy to be the recognition that propositional logic must be superseded by the logic of question and answer[4] is conveniently ignored by his analytic critics when they subject both it and Collingwood's thought in general to the methods of analytic scrutiny.

The purpose of this collection of essays is both to allow Collingwood to restate his published views on political philosophy, an area of his thought that has suffered relative neglect, and to develop, clarify, and illuminate those views by publishing for the first time extracts from the unpublished manuscripts which have a direct bearing upon the problems discussed in the published writings. It is well known that Collingwood was extremely reluctant to leave decisions about the publication of his work to others. On his death he had partially revised *The Idea of Nature* and substantially completed a large proportion of *The Principles of History*, both of which received his authorization for publication, but only the former of which appeared in print, more or less according to his wishes.[5]

---

[3] See e.g. A. Donagan, *The Later Philosophy of R. G. Collingwood* (Oxford, Clarendon Press, 1962). An anonymous reviewer captures perfectly what I have in mind when he or she says: 'If you think of Collingwood as answering badly questions that Professor Ryle has since answered correctly, this will blur your vision. It is not surprising that for Mr Donagan all of Collingwood is "feeble", "confused", "muddled", a "morass of contradictions", "gross errors", "question begging", "howlers" and "tergiversations". This is because the sceptical and "realist" tradition from which the school of Ryle derives is alien to the characteristic features of Collingwood's thought.' Anon. 'Philosopher Without Followers', *Times Literary Supplement* Friday, 26 Apr. 1963.

[4] Collingwood, *Autobiography*, p. 52. He says: 'In logic I am a revolutionary.'

[5] Knox may have exercised more editorial discretion than he admits in the 'Prefatory Note' to *The Idea of Nature* (London, Oxford University Press, 1965: first published posthumously 1945). It appears that he suggested to Collingwood's first wife, Ethel W. Collingwood, and she agreed, that revisions of the section on Whitehead, written in Collingwood's later hand, should be excised from the published version. See the letter from E. W. Collingwood to T. M. Knox, undated. Knox MS 37525/386, University of St Andrews, Scotland. Also see the letter from Ethel W. Collingwood to T. M. Knox dated 22 July 1944 in which she refers to the omission by Knox of a number of manuscript sheets from the end of the work. Knox

The latter was all but rejected, only short extracts being incorporated by Sir Malcolm Knox into *The Idea of History*: Knox disapproved of much that Collingwood wrote in his final years, and as greater accessibility of the manuscripts has led to considered reassessments of their evident quality, it appears that he may have been over-zealous in protecting the intellectual integrity of his mentor's memory.

What justification can there be for publishing items from the unpublished manuscripts? It is clear that a justification needs to be given in the light of Collingwood's documented views on such matters. Ethel Collingwood reports that at least three people who knew Collingwood can testify that he did not wish his manuscripts to be published after his death, and that Bill, Collingwood's son, had read through them all and was of the opinion that his father's reputation should not be marred by allowing any of them to be published.[6] In addition, Kenneth Sisam of the Clarendon Press informed Kathleen, Collingwood's second wife, that her husband had told him that he did not wish any of his work published unless prepared by Collingwood himself for publication.[7] Collingwood wrote to Chapman of the Clarendon Press saying, in a light-hearted manner, that should any exotic disease befall him in the East, the Delegates need have no fear that they would be asked to publish his lectures and other manuscripts, 'or any part of them'.[8] In *An Autobiography* Collingwood suggested that he hated 'leaving a decision of that kind to executors'.[9]

On the other hand Collingwood's will does allow for the publication of material which is not ephemeral or temporary in character, the judges of which are to be the appropriate officers of the Clarendon Press in consultation with his wife.[10] It is clear from a

MS 37524/437. I would like to express my gratitude to Dr James Connelly for bringing the existence of the Knox correspondence to my attention.

[6] Ethel Collingwood in a letter to T. M. Knox dated 22 Aug. 1948. Knox MS 37524/438.

[7] Letter from Kenneth Sisam to K. F. Collingwood, dated 20 Jan. 1943. Clarendon Press Archives, Oxford.

[8] Letter from R. G. Collingwood to the Clarendon Press, dated 14 Nov. 1938. Clarendon Press Archives, file 824121/4662.   [9] Collingwood, *Autobiography*, p. 43.

[10] The original is in the hands of Mrs Teresa Smith, the daughter of Kathleen and Robin Collingwood. Mrs Smith read me the conditions during one of my visits to discuss the prospect of a volume of this kind. It was originally meant to be a companion volume to my *The Social and Political Thought of R. G. Collingwood* (New

reading of the unpublished manuscripts that some items at least were superseded by subsequent rewrites, not because Collingwood thought them inadequate, but because of considerations of a different kind. This is particularly the case in relation to Collingwood's writings on politics. In *The New Leviathan* he makes reference to the 'many thousands of pages of manuscript on every problem of ethics and politics',[11] which form the foundation upon which his book stands. He is quite explicit about the fact that he has been highly selective in choosing the elements discussed and the degree of depth to which they are treated. With reference to the different stages in the argument of *The New Leviathan* he says 'we want to understand only so much as we need in order to understand what is to be said about the next'.[12] It is to be inferred from this that Collingwood failed to include aspects of the manuscripts in *The New Leviathan*, not necessarily because they did not meet his high standards for published work, but because they did not serve the specific purpose he wished to pursue at that time. Indirect evidence for this inference is to be found in the note which precedes the 1929 lectures on moral philosophy. Collingwood, having completely revised the series of lectures, says that the value of the 1929 manuscript 'is that it contains several long passages dealing with various matters in a degree of detail which limits of space denied in the later lectures'.[13] It is reasonable to assume that Collingwood worked on the same principle in relation to *The New Leviathan*.

The unpublished items included in the present collection meet with Collingwood's own high standards, and, in particular, illuminate far more clearly than the cursory discussion in *An Autobiography*, or in the short expositions in *The New Leviathan*, the

York, Cambridge University Press, 1989), but Collingwood's instruction that the Delegates of the Clarendon Press must decide on matters of publication precluded my offering it to Cambridge University Press. The conditions imposed by Colliingwood on the publication of his work are paraphrased by Mrs Kathleen Collingwood in a letter to Kenneth Sisam, 18 Jan. 1943. Clarendon Press Archives. Mrs Smith imposed a further condition that no item should be reproduced in its entirety. The current volume, after protracted negotiations, has met all the conditions.

[11] R. G. Collingwood, *The New Leviathan or Man, Society, Civilization, and Barbarism* (Oxford, Clarendon Press, 1942), p. v.

[12] Ibid., I. 16. Cf. 'NL fasc. 1.2. = Preface', Collingwood MS, DEP 24, pp. 6–7 (this book, chap. 19). References to Collingwood MSS which are followed by DEP and a number (e.g. DEP 7) are held in the Bodleian Library, Oxford University.

[13] R. G. Collingwood, 'Moral Philosophy Lectures [1929]'. Collingwood MS DEP 10, p. iv. The folder clearly marked 1929 which originally housed these lectures is in the possession of Mrs Teresa Smith.

distinctions he wished to make between utility, right, and duty.[14] Collingwood had spent over twenty years developing the terms in which he was to break free of the conventional identity of right and duty, of which even Kant was guilty along with the modern realists G. E. Moore, E. F. Carritt, and W. D. Ross. When Collingwood eventually came to make fully explicit, rather than to intimate, the grounds for his deviation from convention, it was within the context of the grandiose project of *The New Leviathan*. Although Collingwood was never one to be reticent about going against the grain in his opinions, he does appear to have been more cautious than usual about exposing his views on the distinction between right and duty to public scrutiny. This was because of the confidence with which the proponents of the identity, Carritt and Moore, and (less confidently) Ross, expounded their views.[15] Indeed, even when Collingwood made known his inclination to distinguish right from duty, both in published and unpublished papers he would decline to enter into the details of the distinction.[16] For example, in a paper which he read before the Exeter College Dialectical Society, 3 March 1930, Collingwood excuses himself, saying: 'Concerning duty, which I take to be a third form of goodness, I prefer to say nothing here, because I have already trespassed long enough on your patience.'[17] In the moral philosophy lectures of 1929, 1932, 1933, and 1940, Collingwood was much more confident in expounding his views to a student audience, and in articulating them in the course of criticizing the theories of Moore, Carritt, Prichard, and Ross. It is in these lectures, as well as in those of 1921 and 1923, that we see more explicitly than in *The New Leviathan*, the application of a scale of forms analysis to the determination of the philosophical concept of action. Part One of this book introduces the reader to Collingwood's understanding of action in terms of three forms of goodness, or forms of practical reason. Each and every action is the embodiment of the universal categories of utility, right, and duty. It is only in the context of Collingwood's understanding of action in general that his conception of political action as activity according to rule, or

[14] Id., *Autobiography*, pp. 148–9; *New Leviathan*, pp. 99–124.
[15] Collingwood gives this reason in a letter to H. A. Prichard dated 9 Feb. 1933. Correspondence of H. A. Prichard 1925–44, Bodleian Library, Oxford, MS Eng. Lett. d. 166, fo. 34.
[16] See R. G. Collingwood, 'Political Action', *Proceedings of the Aristotelian Society*, 29 (1928–9), 159 (this book, chap. 4).
[17] Id., 'The Good, the Right, and the Useful', Collingwood MS, DEP 6, p. 18.

regularian action, becomes fully intelligible. Politics exhibits its own form of goodness, namely, orderliness, which is achieved by promoting the common good of the community. Punishment facilitates the achievement of orderliness. In this respect the good that it is used to promote is a distinctly political good, neither utilitarian nor moral. In other words, the essence of punishment for Collingwood is neither deterrence nor reform, but instead retribution. At the same time, all concrete actions exhibit the universal characteristics of utility, right, and duty, and therefore deterrence and reform, while not being the essence of punishment, may be counted among its logical properties.

In Part Two of this collection of essays many of the principles which inform Collingwood's analysis of the philosophical concept of action emerge in the context of considering particular practical problems. His understanding of the relationship between philosophy and practical life, or theory and practice, sets the stage for Collingwood's conception of liberalism and the threats to which he thought liberalism was being subjected. Liberalism, for Collingwood, is the political expression of the freedom of consciousness towards which the rational mind strives to develop. The enemies of liberalism, especially Fascism, Nazism, and totalitarian socialism (but not its democratic variants), seek to subvert the development of rational consciousness and deny individual freedom by circumventing individual choice. Liberalism, for Collingwood, is not a political programme, but a method or style of politics. It is the determination to seek beneath the surface of every conflict some fundamental agreement which will facilitate a solution. Liberalism is correlative with the dialectical solution of problems, whereas its enemies practise eristic, or adversarial, styles of politics in which the defeat and suppression of opposition displace any notion of co-operation.

E. W. F. Tomlin has suggested that *The New Leviathan* is not merely a political treatise, because it 'rescues from oblivion'[18] many of Collingwood's ideas of ethics and other matters which otherwise would never have seen the light of day. It is hoped that this collection of essays, as well as adding to our understanding of the place of ethics in the philosophical conception of action as a whole, will reveal the intricacies of aspects of Collingwood's theory of politics which previously have been unknown, or which have not

---

[18] E. W. F. Tomlin, *R. G. Collingwood* (London, Longmans Green, 1961), 32.

been fully appreciated because they have not appeared central, or even marginal, to the main areas of interest in Collingwood studies.

In the remainder of this introduction I wish to provide a context to which each of the items reproduced in this collection can be related, in the hope that the meaning and significance of each will be enhanced. In the first place I wish to explore Collingwood's general political outlook, and have found it convenient to do so by addressing the question of whether his political views underwent a transformation at the time of writing *An Autobiography*. Second, a discussion of Collingwood's conception of philosophy and its implications for the relation between theory and practice will act as a prelude to the determination of the forms of practical reason and their relations to politics. The principal forms of practical reason and their respective forms of goodness are utility, right, and duty, which are characteristics of all action. The goodness of each is correlative with rational choice. Caprice, although choice, is the mere act of choosing without knowing why, or simply being incapable of giving reasons for the choice one has made. Each of the three levels of practical reasons includes within itself a lesser degree of capriciousness than the level which precedes it. The higher the place on the scale each occupies, the greater and more adequately it exhibits the generic essence.

II

It is often remarked that about the time that Collingwood wrote *An Autobiography* his political views had swung sharply to the left, and it is implied, or intimated, that his illness must be held partially responsible for this turn for the worse. Friends and critics alike suspected that he had been converted to Marxism, or had become a Communist sympathizer.[19] The publication of *An Autobiography* was regretted by most of Collingwood's friends, not least because of the

---

[19] See e.g. T. M. Knox, 'Collingwood, Robin George (1889–1943)', *Dictionary of National Biography 1941–1950*, p. 170; A. Quinton, *Thoughts and Thinkers* (London, Duckworth, 1982), 36, 277; H. S. Harris (ed.), *Genesis and Structure of Society* by G. Gentile (Urbana, University of Illinois Press, 1960), introduction, p. 20; R. H. S. Crossman, 'When Lightning Struck the Ivory Tower', in *The Charm of Politics* (London, Hamish Hamilton, 1958), 108; H. J. Laski, review of *New Leviathan* in *New Statesman and Nation*, 24 (1942), 98; S. Toulmin, 'Conceptual Change and the Problem of Relativity' in M. Krausz (ed.), *Critical Essays on the Philosophy of R. G. Collingwood* (Oxford, Oxford University Press, 1972), 219.

apparent change in his political views.[20] C. C. J. Webb, the historian and idealist theologian, had heard in February 1939 that his friend, R. G. Collingwood, had declared himself a Marxist convert. When Collingwood returned, in April 1939, from his long recuperative cruise to the East Indies,[21] sporting a beard, substance was added to the rumour.[22] However, the strident tones of *An Autobiography*, considered by one commentator to contain 'a good deal of the "I'll-tell-the-world" spirit',[23] made it sufficiently ambiguous to allow diametrically opposed interpretations. Howard Hannay, for example, contends that 'Mr Collingwood does not agree with Marx any more than with Fascism'.[24] On the other hand, Joseph Needham felt able, on the same evidence, to invoke Collingwood as one of a growing number of intellectuals in Britain who see 'in the Marxist world-outlook the only approach to a satisfactory philosophy of history'.[25] It is ironic that readers of the *Autobiography* appear to

[20] R. B. McCallum, 'Robin George Collingwood 1889–1943', *Proceedings of the British Academy*, 29 (1943), 467.

[21] For details of the journey see R. G. Collingwood, 'Log of a Journey in the East Indies: 1938–9', DEP 22. The Bodleian Library holds a photocopy of the original which Mrs Teresa Smith has in her possession.

[22] C. C. J. Webb, journal entries 18 Feb. and 2 May 1939, Bodleian Library, Oxford, MS Eng. Misc. e. 1175. I owe this example to J. Patrick's eminently readable and scholarly study of *The Magdalen Metaphysicals: Idealism and Orthodoxy at Oxford 1901–1945* (Atlanta, Mercer University Press, 1985), 159.

[23] SLP, review of *Autobiography* in *Journal of Philosophy*, 35 (1939), 717.

[24] A. Howard Hannay, review of *Autobiography* in *International Journal of Ethics* 51 (1941), 370. I hope the reader will allow me a digression. Hannay was the co-translator, with Collingwood, of G. de Ruggiero, *Modern Philosophy* (London, Allen and Unwin, 1921). Hannay had translated the work much earlier and submitted it under the title of *Contemporary Philosophy* to Macmillan in 1914. On the whole Macmillan's reader, Sir Henry Jones, was extremely enthusiastic about de Ruggiero's book, but had two reservations: first de Ruggiero said too little of Bosanquet, and wrongly tried to relate Green and McTaggart, and second the translation, although conscientious, was stiff and clumsy. See Jones, reader's report on *Contemporary Philosophy* by de Ruggiero, trans. A. H. Hannay. Macmillan Archives, Vol. mcciii, 1911–21, fos. 64–5, Brit. Mus. Lib. MS 5598. It appears that Macmillan may have declined to publish it because of the war (*Modern Philosophy*, p. 6), and Hannay then enlisted Collingwood's help with the translation. It is interesting that the 'Translators' Preface' of *Modern Philosophy* addresses Jones's first reservation by suggesting that philosophical developments have occurred since it was first published, namely Bosanquet's Gifford Lectures and the new prominence of British and American realism; but on the grounds that no history can remain up to date for ever, and to avoid impairing the overall unity of the book, the translators declined to ask the author to insert a new paragraph here and there (*Modern Philosophy*, pp. 6–7). Collingwood refers to this translation in 'List of Work Done', Collingwood MS, DEP 22, p. 21: 'Ruggiero's Fil. Contemp. (from first draft by A. H. Hannay) 1919–20'.

[25] Needham, letter to *The New Statesman and Nation*, 10 Feb. 1940, p. 175.

have exercised a similar lack of care in determining the position it wished to convey as that of which Collingwood complained in rebuking the 'realist' school for misrepresenting the philosophers whom it criticized.[26]

What evidence is there for believing that Collingwood's political views became suddenly more left wing in character during the late 1930s? It is appropriate to look at *An Autobiography*, for this is the main source of the contention, and compare the views held there with those written before and after its composition. In this respect it is convenient to distinguish Collingwood's attitudes to liberalism, Marx, and socialism.

In *An Autobiography* Collingwood contends that his politics had always been liberal in the continental sense, or what the British call democratic. For him this entailed, in the British context, representative democracy in which each citizen had an obligation to vote based upon a sound political education, facilitated by the dissemination of a wide range of views through a free press, and by 'a universally recognized right of free speech'.[27] From early in his career Collingwood had subscribed to liberal principles. He greatly admired the Italian philosopher and liberal Guido de Ruggiero whose social liberalism, in the tradition of Green and Hobhouse, Collingwood found particularly attractive. Collingwood translated Ruggiero's *Modern Philosophy*, with A. H. Hannay, during late 1919 and early 1920. He also translated, for his own pleasure, Ruggiero's *Scienza*, in 1920, after which he wrote to the author exclaiming: 'in you I for the first time find and possess myself'.[28] In the Michaelmas term of the same year Collingwood offered a series of lectures on de Ruggiero's philosophy,[29] and after the Italian published *Storia del liberalismo europea*[30] Collingwood wrote to Kenneth Sisam, of Oxford University Press, urging its translation and publication in English. Collingwood enthusiastically recommends de Ruggiero as one of the most brilliant of Italian historians and philosophers, and certainly the most important of the post-Crocean generation.[31] During the course of translating the book Collingwood wrote to de

---

[26] Collingwood, *Autobiography*, pp. 22–3.   [27] Ibid. 153.
[28] Letter to de Ruggiero dated 1 July 1920, DEP 27.
[29] Collingwood, 'List of Work Done', p. 51.   [30] Bari: Literza, 1925.
[31] R. G. Collingwood, letter to Kenneth Sisam dated 1 Aug. 1925. Clarendon Press Archives, file 5898, C7500.

Ruggiero, 2 September 1926, declaring that: 'The political principles expounded and implied are at every point my own, and expressed with a justness and completeness that leave me nothing to do but express my complete agreement.'[32]

The primary principle upon which liberalism rests is, for de Ruggiero, the repudiation of all reductionist interpretations of human action, and the assertion of spiritual freedom in which individual consciousness and the expression of the self are allowed to develop unimpeded.[33] Only the person who is free, that is an autonomous intelligent agent, can acknowledge and recognize the possession of freedom by others.[34] The human mind, in the pure act of thinking, creates the experience of which it is conscious. De Ruggiero's 'actualism', unlike that of Giovanni Gentile, sees the spiritual development of the individual, not in terms of dominating others and subordinating them to his or her will, but in terms of the liberation of the self from internal passions and external constraint. This does not mean that freedom in the form of caprice or licence, which promotes the actions of one person while impairing those of others, can be permitted, because it would negate the mutual recognition of freedom and lead to the destruction of civil life.[35] On the contrary 'The really free man is not the man who can choose any line of conduct indifferently—this being rather a frivolous and weak-willed man—but the man who has the energy to choose that which is more conformable to his moral destiny.'[36] Freedom, then, is acting in accordance with one's duty, and it was Kant, de Ruggiero contends, who was responsible for illuminating this truth.

The primary principle of liberalism, that is, the celebration of rational freedom, is what Collingwood spent a great deal of his life attempting to articulate. The moral philosophy lectures, and all the other associated writings on economics, politics, and ethics, emphasize that freedom is a matter of degree, the highest practical form of which is manifest in doing one's duty.[37] We create the world in our

---

[32] DEP 27. It was to de Ruggiero, of course, that Collingwood sent the only extant copy of *Libellus de Generatione: Autobiography*, p. 99.

[33] De Ruggiero, *European Liberalism*, p. 357, and 'Liberalism', *Encyclopaedia of the Social Sciences*, ix (London, Macmillan, 1933), 435.

[34] Id., *European Liberalism* p. 357.   [35] Ibid. 356.   [36] Ibid. 351.

[37] For a bibliography, which includes the political and related writings, see D. S. Taylor, 'A Bibliography of the Publications and Manuscripts of R. G. Collingwood, with Selective Annotation', *History and Theory*, 24 (1985). This bibliography needs to be supplemented by P. Johnson, 'The Letters of R. G. Collingwood: An Initial

acts of thought and the less clear we are in our thinking the more confused our world of experience. Like Lachelier, the French idealist, Collingwood equates freedom with knowledge.[38] Such knowledge was for Collingwood historical, and *The Principles of History*, only partially completed at his death, was to demonstrate the relation between freedom and historical knowledge; that is, to extol the virtues of a historical as opposed to a scientific civilization.[39] *The New Leviathan* demonstrated the identity of freedom and knowledge from a different perspective: the *Principles of History* was to explore freedom in relation to the highest level of theoretical consciousness, whereas his political treatise explored it in relation to the highest level of practical consciousness, namely duty.

Like de Ruggiero, Collingwood believes that consciousness of one's freedom necessarily entails the recognition of the freedom of others. To be conscious of one's freedom, and to be socially conscious are correlative. Society depends upon the mutual recognition of freedom. In a sermon preached in 1935 Collingwood argued that each mentally mature person:

Descriptive Checklist', 1975, unpublished. Johnson, of the University of Southampton, has uncovered numerous letters, and since writing the checklist has discovered many more: J. M. Connolly, 'Metaphysics, Method and Politics' (Ph.D. dissertation University of Southampton, 1984), app. 3, 4, pp. 411–92. J. Patrick's *Magdalen Metaphysicals* brings to light many of Collingwood's letters and other writings which are not listed elsewhere. Of particular interest are Collingwood's letters to J. A. Smith (J. A. Smith Papers, Magdalen College Library, Oxford), and an essay entitled 'The Theory of History' which Patrick claims is likely to have been authored by Collingwood. See *Magdalen Metaphysicals*, p. 56.

There are a number of other items not listed elsewhere which may be of interest to the Collingwood scholar: 'Luxemburg', *Encyclopaedia Britannica*, 12th edition (1920) (Collingwood refers to this in his 'List of Work Done', p. 23); also his translation of 'Science, History and Philosophy' by de Ruggiero, *Journal of Philosophical Studies*, 6 (1931). There is a need to add to the unpublished manuscripts 'An Essay on F. H. Bradley's *Appearance and Reality*' (1932), and the letter of application for the Waynflete Chair of Metaphysical Philosophy, which includes a testimonial written by S. Alexander. Both unpublished items are in the possession of Mrs Smith. In addition, Collingwood's letters to Knox, University of St Andrews Library, Scotland, are not listed by Taylor.

[38] See R. G. Collingwood, letter to T. M. Knox, 2 Nov. 1937. Knox MS, University of St Andrews Library, MS 37524/421. See *The Philosophy of Jules Lachelier*, translated and introduced by E. G. Ballard (The Hague, Nijhoff, 1960), 118.

[39] See van der Dussen, *History as a Science*, App. 1, 'Scheme for The Principles of History', pp. 431–2. Items 10 and 16 of this collection indicate to some extent what Collingwood understood by historical and scientific civilizations.

is bound to recognize in others the same kind of freedom which he claims for himself; and therefore each must recognize the will of these others as a source of rules binding on himself. And this is not a mutual slavery, because the joint enterprises out of which these rules grow are pursued by each party of his own free will.[40]

In *The New Leviathan* Collingwood reiterates the primary principle of liberalism in saying that

social consciousness involves the consciousness of freedom. A society consists of persons who are free and know themselves to be free. Each knows the others to be free as well as himself. If consciousness of freedom is a mark of being mentally adult, a society can consist only of mentally adult persons.[41]

In other words freedom is not something with which we are born; it is an achievement that signifies mental maturity. This entails a second principle of liberalism to which de Ruggiero and Collingwood subscribe, namely, the idea that freedom can only be developed and attained within the context of an appropriately constituted community whose institutions are designed to facilitate the progression of consciousness towards liberty.[42] The general aims of education, in Collingwood's view, are to provide the impetus to human development, making the pupil better able to face up to the demands of life.[43] Education means 'helping a mind to create itself'[44] and thus reach a level of consciousness requisite to full participation in social life. It is the process which encourages self-control of our lower natures,[45] and the necessary expression of our emotions, which if suppressed entail a denial of our emotional life that leads to the transference to the higher levels of consciousness of false information upon which intellect builds a distorted edifice. This is what Collingwood calls the 'corruption of consciousness'.[46] Education, essentially, is the means by which the heritage of a civilization

[40] R. G. Collingwood, 'Rule-making and Rule-breaking', sermon preached in St Mary the Virgin, Oxford, 5 May 1935, Collingwood MS, DEP 1 p. 10.

[41] Id., *New Leviathan*, 20.23.

[42] De Ruggiero, *European Liberalism*, pp. 32, 355; Collingwood, *New Leviathan*, 33.18.

[43] R. G. Collingwood, 'The Place of Art in Education', *Hibbert Journal* 24 (1926), 435, 443.

[44] Id., *Speculum Mentis* (Oxford, Oxford University Press, 1924), 316.

[45] Id., 'Science, Religion, and Civilization', the third of a series of lectures under that title delivered in Coventry Cathedral, Oct.–Dec. 1930, by J. Needham, B. J. Streeter, and R. G. Collingwood. DEP 1, pp. 1–2.

[46] Id., *The Principles of Art* (Oxford, Oxford University Press, 1977: first published, 1938), 217–20, 251, 282–5, 336.

is passed on to a new generation of actual and potential contributors to social life.[47]

In the political sphere liberalism implies that 'political activity and political education are inseparable, if not identical'.[48] For Collingwood, 'the life of politics is the life of political education'.[49] The liberal, or democratic, system is, in Collingwood's view, 'not only a form of government but a school of political experience coextensive with the nation', whose strength lies in 'a politically educated public opinion'.[50] The importance of the role of political education, promoted by those who rule, is reiterated and emphasized in both 'The Three Laws of Politics' and *The New Leviathan*, written after *An Autobiography*. In the former Collingwood contends that it is the business of rulers to conduct themselves in ways which constitute examples to be emulated.[51] In the latter it is argued that the social element or ruling class in a community has the responsibility of ensuring the provision of the necessary political education to facilitate the conversion of the non-social, that is the mentally and politically immature, to membership of the social community.

A further essential of liberalism, in Collingwood's view, is the resolution of conflict by dialectical means. Opinions circulate freely, informing the life of a society whose politicians are constantly exposed to public scrutiny and criticism. The dialectical method associated with liberalism entails the presentation of differing views which are the object of free discussion. Eventually, out of the opposition emerges 'some common ground on which to act'.[52] Contrary to what one may think, this is not the politics of compromise: it is the recognition that beyond the good of each group or individual associated with a particular opinion, is a common good, or the good of society as a whole, the primary principle of which is that the continuing existence of that society be ensured.[53]

These principles of liberalism (in its European sense) permeate

---

[47] Id., *New Leviathan*, 39.18. A delicate and sensitive job for which professional educators are ill equipped, and parents admirably suited.

[48] Id., 'Man Goes Mad', rough MS begun 30 Aug. 1936. Collingwood MS, DEP 24, p. 17 (this book, chap. 14).

[49] Id., *New Leviathan*, 32.34. Cf. 27.12, 27.38, 32.31.

[50] Id., *Autobiography*, p. 153. Cf. id., *An Essay on Metaphysics* (Oxford, Oxford University Press, 1940), 134.

[51] Id., 'The Three Laws of Politics', p. 10. (This book, chap. 18).

[52] Id., 'Man Goes Mad', p. 16. (This book, chap. 14).

[53] Id., 'The Breakdown of Liberalism', Collingwood MS, DEP 17, pp. 1–2, undated. Cf. *New Leviathan*, 24.57, 24.6.

the whole of Collingwood's political writings both before and after the composition of *An Autobiography*. The social arrangements which allow these principles to operate are actually enunciated in what Collingwood calls the three laws of politics. First, every society has a ruling and a ruled element. The former takes the initiative, formulates solutions, and implements them, while the latter is relatively devoid of initiative and does what the rulers tell it.[54] Second, the division between the rulers and the ruled must be permeable, and adequate provision must be made for educating a constant stream of aspirant converts.[55] The third law of politics maintains that the ruled element has a tendency to imitate the rulers, and thus is behoves the rulers to provide worthy examples to follow.[56] However, it is true to say that Collingwood wished to emphasize that a society in which the three laws of politics operate is somehow different from the liberal conception of society. Liberalism, he suggests, in championing freedom, only brings into view one connotation of what a body politic entails, and neglects the opposite connotation which is also present, that is, the existence of a ruled element the members of which are subordinate to, and follow the initiatives of, the rulers who are the stronger element.[57] Collingwood seems to be suggesting here that liberalism entails a belief in equality. The equality of which continental liberalism spoke was an equality of opportunity which is entirely compatible with the idea of a ruled and ruling class as long as the former is afforded equal opportunities to develop the consciousness of freedom to enter the latter.[58] Croce, for example, believed that liberalism did not imply equality, but, on the contrary, promoted an aristocracy that was not exclusive of democracy.[59] He argued that 'aristocracy is truly vigorous and serious when it is not a closed but an open aristocracy, firm in rejecting the crowd, but always ready to

---

[54] Id., *New Leviathan*, 25.7, and 'Three Laws of Politics', pp. 5, 9 (this book, chap. 18).

[55] Id., *New Leviathan*, 25.8., and 'Three Laws of Politics', p. 8 (this book, chap. 18).

[56] Id., *New Leviathan*, 25.9., and 'Three Laws of Politics', p. 10 (this book, chap. 18).

[57] Id., 'The Three Laws of Politics', p. 13. (this book, chap. 18).

[58] De Ruggiero, *European Liberalism* p. 437.

[59] B. Croce, *Philosophy of the Practical: Economic and Ethic*, trans. D. Ainslie (New York, Biblo and Tannen, 1967), 132.

welcome those who have raised themselves to its level'.[60] Colling-wood's own view of European liberalism is at variance with his criticism of liberalism's one-sidedness. In the Preface to de Ruggi-ero's *European Liberalism* Collingwood contends that liberalism is neither democratic nor authoritarian, but partakes of both doctrines. It is 'democratic in its respect for human liberty, it is authoritarian in the importance it attaches to necessity for skilful and practised government'.[61] Collingwood's analysis, in *The New Leviathan*, of the necessity of the coexistence of the principles of democracy and aristocracy in the body politic is an elaboration of Croce's view,[62] and in this respect demonstrates the Englishman's continuing allegiance to continental liberalism.

Shortly after writing *An Autobiography* Collingwood extols the virtues of liberal principles,[63] and says, quite explicitly, that 'what liberals or democrats think, there can be no doubt of it, is wise'.[64] If Collingwood did take a turn to the left in *An Autobiography* it must have been very short-lived because he was writing a defence of the traditional values of European civilization[65] almost simultaneously with correcting the proofs and rewriting the last chapter of *An Autobiography*.

What needs to be emphasized is that, like the Italian liberals Croce and de Ruggiero, Collingwood equated liberalism with free-dom, society and civilization. All four concepts are inextricably related. Freedom is correlative with the higher levels of conscious-ness, and it is upon the recognition of the freedom of others that social life depends. The socializing process encourages the conduct of social relations in terms of a dialectical, rather than an eristical, or adversarial approach to problems. It is the process of working towards the gradual elimination of force from human relations, an ideal which can only ever be approximated because of the permanent existence of a non-social element in the body politic which is

---

[60] Id., *Politics and Morals*, trans. S. J. Castiglone (London, Allen and Unwin, 1945), 83. Richard Bellamy has done much in recent years to revise interest in Italian social theory. See especially *Modern Italian Social Theory: Ideology and Politics from Pareto to the Present* (Cambridge, Polity Press, 1987). I have dealt with Collingwood's relation to Italian and British idealists more comprehensively in chapters 1 and 5 of my *Social and Political Thought of R. G. Collingwood*.
[61] Collingwood, translator's preface, p. viii (this book, chap. 13).
[62] Id., *The New Leviathan*, chap. 26.
[63] Id., *Essays on Metaphysics*, pp. 133–42.
[64] Id., 'Fascism and Nazism', p. 173 (this book, chap. 15).
[65] Id., *Essays on Metaphysics*, pp. 133–42.

dominated by the stronger wills of the social. The non-social can never be eliminated because it is constantly being replenished with new additions (i.e. criminals who have rejected society and children who have not yet attained the level of consciousness necessary for admittance to the social element). The socializing process, then, is similar to the Kantian categorical imperative of treating each person as an end in herself or himself, rather than as a means to an end. Civilization is the exact equivalent of socialization in that it is the process which encourages the development of the ideal of civility, namely the gradual elimination of force from our relations with members of our own and other societies. In addition, civilization is the development of a more sensitive attitude towards our relationship with nature, encouraging intelligent exploitation, rather than ruthless decimation, of the earth.

All through his career Collingwood was concerned about the ways in which European civilization, and by implication the liberal ideal of freedom, was being threatened and undermined from all sides. Liberalism, he acknowledged, had not been a wholehearted success,[66] but its values were worth preserving, whether against the threat of the Prussian philosophy;[67] the disintegration of the unity of life of the mind;[68] the 'time of crisis and chaos in philosophy';[69] the encroachment of the town upon the countryside and the consequences of the industrial revolution;[70] the debasement and devaluation of art;[71] the pernicious effect of positivism and realism, and their associated utilitarian civilization;[72] the failure of the popular press to uphold democratic values, and the actions of a

[66] Id., 'Man Goes Mad', pp. 26–7 (this book, chap. 14).

[67] Id., 'The Spiritual Basis of Reconstruction', address to the Belgian Students' Conference at Fladbury, 10 May 1919. Collingwood MS, DEP 24, pp. 8–16 (this book, chap. 17).

[68] Id., *Speculum Mentis*, p. 30. Collingwood says: 'What is wrong with us is precisely the detachment of these forms of experience—art, religion, and the rest— from one another; and our cure can only be their reunion in a complete and undivided life,' p. 36. What this means is that we will recapture the freedom of the middle ages; 'the freedom of occupying an ordained place which one desires to occupy and finds happiness in occupying.'

[69] Id., *An Essay on Philosophical Method* (Oxford, Oxford University Press, 1933), 6.

[70] Id., 'Man Goes Mad', pp. 29–37, and 'Art and the Machine', Collingwood MS, DEP 25, pp. 11–12. Probably written in the early 1930s.

[71] Id., 'Art and the Machine', pp. 1–10, and *Principles of Art*, pp. 78–104.

[72] Id., *Autobiography*, pp. 44–52, and [Fairy Tales] 'IV. Magic', Collingwood MS, DEP 21, pp. 14–19 (this book, chap. 16).

deceitful and corrupt press;[73] the threat of irrationalism to the traditional value of seeking after truth;[74] and finally, in the face of threats from the left and right, and particularly Fascism and Nazism which Collingwood viewed as the New Barbarism.[75] In essence, Collingwood thought that civilization, and its consequent ideals, was besieged by hostile forces. *An Autobiography* simply reflects this growing concern, which had gradually become accentuated during the whole of the 1930s. It constitutes a more public, and more radical, pronouncement of his concerns, but it does not constitute a rejection of the values of liberalism in favour of those of another creed. Indeed, he was profoundly dismayed that liberalism had become complacent, and appeared to lack the vitality and energy to defeat its enemies, a view which he elaborated upon in 'Fascism and Nazism'.[76]

These, then, are the principles of liberalism to which Collingwood subscribed, and from which the *Autobiography* detracts little or nothing. His adherence to these principles betrays no radical shift to the left. If it was not the principles that betrayed Collingwood's move to the left, perhaps it was his stand on particular issues. The most obvious candidate is his position on Appeasement which seems to have distressed many of Collingwood's colleagues, and presumably for them constituted evidence of his move to the left.[77] It is true, of course, that prior to the Nazi–Soviet Pact (23 August 1939) the Communist Party and Marxists in general were opposed to the policy of Appeasement. The policy demonstrated, for them, the Fascist sympathies of the Chamberlain Government. It was the 'National' Government's betrayal of Spain, Abyssinia, and Czechoslovakia that confirmed for Collingwood the anti-democratic tendencies of its leaders. Indeed, he believed that the engineering of a war-scare during the Munich crisis fell firmly within the tradition of the methods of Fascist dictatorship. His views, in this respect, do not necessarily constitute evidence of a shift to the left. The Appeasement issue split the country and largely transcended partisan loyalties.

The Oxford By-election of 1938 conveniently focuses the issue,

[73] Id., *Autobiography*, pp. 155–67.   [74] Id., *Essay on Metaphysics*, pp. 133–42.
[75] Id., 'Man Goes Mad', pp. 19–26 (this book, chap. 14); *Autobiography*, pp. 157–67; 'Fascism and Nazism', pp. 168–76; and *New Leviathan*, 45.1–96. I do not wish to imply that the list of threats to civilization given here is exhaustive.
[76] Id., 'Fascism and Nazism', pp. 171–4.
[77] Tomlin, 'R. G. Collingwood', p. 32.

and demonstrates the diversity of party and ideological affiliation with which Collingwood was associating himself in taking a stand against Appeasement. The Labour and Liberal candidates,[78] after a certain degree of disquiet, agreed to stand down in favour of the independent candidature of A. D. Lindsay, whom, it was thought, would serve to unite the opposition against Chamberlain's foreign policy and defeat Quinton Hogg the Conservative candidate. Malcolm Knox has suggested that the strength of Collingwood's feelings about the policy of Appeasement impelled the Oxford don to become actively engaged in politics by going to Labour Party headquarters and entreating the leaders to oppose wholeheartedly the policy of Appeasement.[79] Collingwood was unable to participate in the Oxford By-election because he was recuperating after a serious stroke. He did, however, send a letter of support in which he expressed his concern about the present situation and desire that the democratic spirit would prevail:

I am leaving for the East tomorrow but I cannot go without sending you my deepest good wishes for your success in your candidature. I do not think that the country has ever in all its history passed through a graver crisis than that in which it is now involved. I am appalled by the apathy with which our situation is regarded by a great many of us, and by the success which the Government has had in keeping the country as a whole from knowing the truth. Your candidature shows that the spirit of English democracy is not extinct. I hope that it still survives among those who have to vote next week.[80]

The company in which Collingwood found himself, by supporting Lindsay, was remarkably diverse: for example three future Conservative prime ministers, Winston Churchill, Harold Macmillan, and Edward Heath pledged their support. There were also many eminent Liberals opposed to Appeasement and anxious to see Lindsay win, among whom were Lord Crewe, Sir Archibald Sinclair, and

[78] The candidates were Patrick Gordon Walker for Labour, and Ivor Davies for the Liberals. Walker initially objected most vociferously, but retreated when it was obvious he would not retain the support of the National Executive.

[79] T. M. Knox, review of W. M. Johnston, *The Formative Years of R. G. Collingwood* in *Philosophical Quarterly*, 19 (1969), 166.

[80] Letter from R. G. Collingwood to A. D. Lindsay dated Oct. 20 1938. Lindsay Papers, University Library, Keele. The extract reproduced here is published in D. Scott, *A. D. Lindsay: A Biography* (Oxford, Basil Blackwell, 1971), 251. I am indebted to Johnson, 'The Letters of R. G. Collingwood', for alerting me to both the letter and Lindsay's biography.

Lady Violet Bonham Carter. R. H. S. Crossman, G. D. H. Cole, and H. N. Brailsford are probably the best known of Lindsay's advocates from the Labour camp. Furthermore, the Oxford Anarchists supported Lindsay, in a convoluted way, by pledging to dissuade only Hogg supporters from casting their vote in the election. Numerous academics actively participated in furthering Lindsay's cause, including J. L. Austin who took it upon himself to heckle every evening at Hogg's meetings.

This was a time when emotions were high, and allegations of dubious political affiliations were hurled freely with little or no regard for evidence. Patrick Gordon Walker, for example, repeating allegations made in the *Daily Herald*, suggested that an alliance of intellectuals and Communists initiated the Lindsay campaign and persuaded the local Labour Party of its merits. To which G. D. H. Cole immediately responded by denying that the Communists had anything to do with initiating Lindsay's candidature, and that those in the Oxford Labour Party who were sympathetic were mainly of the moderate wing. Cole claims that he met no Communists either during the preliminary discussions or during the campaign itself. N. H. Brailsford responded two weeks later by suggesting that the Fascist Axis could only be resisted with Russian support, and that it was therefore necessary for the Labour Party to extend a hand to both Liberals and Communists alike.[81] Furthermore, Hogg supporters were suggesting that Stalin had sent telegrams of support to Lindsay, while the Lindsay camp repeated the slogan 'Hitler Wants Hogg'. The rumours which circulated about Collingwood's political leanings have to be viewed in the context of the heightened tensions and suspicions aroused by the dramatic events in Europe and their anticipated implications for Britain. The fact that his newly grown beard was taken as confirmation of his political leanings is indicative of the times.

The claim that Collingwood's politics shifted to the left in the late 1930s is further discredited by the fact that his opposition to Appeasement was perfectly consistent with views that he had held

[81] P. Gordon Walker, 'The Oxford Election', *The New Statesman and Nation*, 5 Nov. 1938, pp. 719–20; G. D. H. Cole, 'The Oxford Election', ibid., 12 Nov. 1938, pp. 756–66. H. N. Brailsford, 'Labour Policy', ibid., 26 Nov. 1938, p. 868. I am indebted to David Blazer of Sydney University of Technology for supplying me with copies of the above correspondence. The best and fullest account of the Oxford By-election appears in chap. 14 of Scott, *A. D. Lindsay*, on which I have freely drawn.

since at least 1919. In the international community of bodies politic, Collingwood argues, states may exhibit differing degrees of civilization. Some will have reached that level of rational consciousness which promotes the establishment of social relations among bodies politic, while others will have attained insufficient maturity to do so. In certain circumstances it may be appropriate for the higher to impose its civilization on the lower. Africa and Asia, Collingwood suggests, may benefit from European imperialistic rule to a similar extent to which Europe benefitted from 'the discipline of Roman rule and the legacy of Roman law'.[82] It should not be assumed that Collingwood was here referring to a crude assimilation of the lower into the higher. He believed that a process of diffusion and fusion occurs, exemplified by the example of the Roman occupation of Europe.[83] Nor is he suggesting that there is any close analogy to be made between the Roman Empire and modern imperialism: 'The empires of modern times are rent by a racial cleavage between a governing race and a governed, which are too far apart to unite into a single whole. We have barriers of colour and race and language which were absolutely unknown in the Roman world.'[84] Collingwood's remarks in the 'Spiritual Basis of Reconstruction' should be taken as an endorsement of a qualified imperialism whose interventions in the affairs of other nations, like those of the state within the nation, are justified on the grounds that they enhance liberty and the consciousness of freedom.[85] This is what Collingwood calls the 'right imperialism'.[86]

The wrong kind of imperialism occurs when one civilized country attempts to impose its will upon another of a similar standard of

[82] Collingwood, 'The Spiritual Basis of Reconstruction', p. 8 (this book, chap. 17).

[83] See the first and second editions of R. G. Collingwood, *Roman Britain* (Oxford, Oxford University Press, 1923, 1932). For an account of the slightly different analyses Collingwood gave of the Romanization of Britain in each edition see van der Dussen, *History as a Science*, pp. 241–7. Collingwood says elsewhere e.g. 'The English character blends the law-abiding Roman's love of sound government with the self-reliance of the seafaring Saxon, and is unintelligible and unmanageable except by people who realize this fact. So close do the lessons of ancient history stand to the problems of modern life.' 'The Roman Signal Station on Castle Hill, Scarborough' (Scarborough, Scarborough Corporation, 1925), 1.

[84] Id., *Roman Britain*, 2nd edn., p. 15.

[85] See A. Vincent and R. Plant, *Philosophy, Politics and Citizenship* (Oxford, Basil Blackwell, 1984), 87–90 for a brief discussion of imperialism and idealism. Cf. Collingwood, 'The Spiritual Basis of Reconstruction', pp. 13–14 (this book, chap. 17).

[86] Collingwood, 'The Spiritual Basis of Reconstruction', p. 8 (this book, chap. 17).

civilization. The immediate cause of the First World War, in Collingwood's view, was an instance of the wrong kind of imperialism. Germany's attempted imposition of its civilization upon other civilized states, while constituting the immediate cause of the war, was nevertheless the manifestation of a more fundamental cause. This cause was the Prussian philosophy which endowed the state with all creative and innovative initiative. The state was answerable only to itself and possessed absolute rights while exercising absolute power over all its subjects. Unless this philosophy was discredited and destroyed, even though Germany had been defeated, it would, in Collingwood's view, emerge again in a more menacing form. Collingwood's views were totally vindicated with the re-emergence of militarism in the 1930s, and with the rise of Fascism and Nazism. Collingwood was completely hostile to pacifism because he believed that far from averting war, pacifists actually encouraged it.[87] Instead, he supported the Thomistic idea of a just war, based upon the principles of 'just cause' and 'right intention'. Collingwood could not accept 'the distinction between offensive and defensive war, which is fatally popular nowadays—fatally because every belligerent can always claim with some show of justice, that he is acting in self-defence'.[88] Every society, Collingwood argued, must be prepared to use force against those external enemies who would attempt to destroy its achievements without replacing them with achievements of a higher value.[89] War can actually promote peace and is perfectly justified if it is the only means by which the members of a body politic under the influence of irrational emotions and subject to a popular ruler who is intent upon aggressive rather than dialectical conduct, can be prevented from extending the tyranny to which they are subject to the bodies politic they wish to subvert.[90]

It now remains to look briefly at Collingwood's attitudes to Marx and socialism to see if evidence for his 'late acquisition of enthusiasm for Marx'[91] can be discerned. Was Collingwood telling the truth when he said that part of him always cheered, at least softly, when he read Marx even though he was never convinced by the economics

---

[87] Id., 'Action: A course of lectures (16 lectures) on moral philosophy' (1923), Collingwood MS, DEP 3, p. 71; 'War in its Relation to Christian Ethics with Special Reference to the Lambeth Report, 1930' (1932), Collingwood MS, DEP 1, p. 4; *New Leviathan*, 29.88–98.

[88] Id., 'War in its Relation to Christian Ethics', p. 3. Cf. *New Leviathan*, 29.86.

[89] *New Leviathan*, 30.88.   [90] Ibid. 30.99.

[91] Quinton, *Thoughts and Thinkers*, p. 36.

and metaphysics?[92] Collingwood actually says very little about Marx prior to writing *An Autobiography* just as he had said very little about Hobbes prior to writing *The New Leviathan*.[93] Collingwood suggests, for example, that as an imperative reminder of the importance of focusing upon economic history Marx's interpretation of events was, in its day, valuable and legitimate.[94] As an economist Marx had the merit of resolving the problems of economics into historical problems,[95] and 'the practical consequences of his teaching have been, perhaps, the most important feature in general politics in the last hundred years'.[96] Nevertheless, Marx's doctrines are philosophically inadequate in abstracting economic facts and presenting them as fundamentally important in history; in conceiving history in too scientific, or positivistic, a manner; and even though Marx has been politically important 'it does not follow that his philosophy of history is likely to satisfy a critical historian'.[97]

Soon after writing *An Autobiography* Collingwood reiterated, in stronger terms, some of the strengths which he believed Marx to possess. Essentially, he saw Marx as having built upon the foundations which Hegel had laid. In *An Essay on Metaphysics* Collingwood contends that Hegel's emphasis upon the study of internal strains was an influence for the better in historical studies. The fact that Marx developed this emphasis in a brilliant analysis of the internal strains inherent in the economic society of the nineteenth century entitles him 'to the name of a great historian'.[98] At about the same time Collingwood contended that Marx was a great man who was 'exceptionally strong' in economic history, and who shared both the strength and weakness of Hegel's philosophy: 'its strength, in penetrating behind the facts to the logical nexus of underlying concepts; its weakness, in selecting one aspect of human life (in Hegel the political, in Marx the economic) as in this sense fully

[92] Collingwood, *Autobiography*, p. 152.

[93] See my 'The Two *Leviathans*: R. G. Collingwood and Thomas Hobbes', *Political Studies* 35 (1987), 443–60.

[94] R. G. Collingwood, 'The Nature and Aims of a Philosophy of History' reprinted in W. Debbins (ed.), *Essays in the Philosophy of History* (Austin, University of Texas Press, 1965), 40.   [95] Id., *Speculum Mentis*, p. 52.

[96] Id., 'Lectures on Philosophy of History', ii (Trinity Term 1929), Collingwood MS, DEP 12, p. 20.

[97] Id., 'The Nature and Aims of a Philosophy of History', p. 40; *Speculum Mentis*, p. 286; 'Lectures on Philosophy of History', p. 20.

[98] Id., *Essays on Metaphysics*, p. 75.

rational in itself'.[99] The fact that Collingwood believed both Hegel and Marx to have considerable strengths did not blind him to their considerable faults. Far from being philosophers of freedom they were, in Collingwood's view, philosophers of servility, of herd worship. The German political experience in which there was no conception of the spontaneous creation of a social order which secured for itself rights against the encroachment of princely power, predisposed German philosophers to be impervious to the idea of society as an 'artificiall Man'. It is Collingwood's contention that both Hegel and Marx denied 'the existence, even the possibility of free joint activity'.[100] This was partly because they had no experience of it. In essence, Collingwood always admired aspects of Marx's work, but never ceased to have serious reservations.

Idealist liberals were not indiscriminatingly hostile to socialism. On the whole, they acknowledged, in varying degrees, the efficacious implications that socialism had upon liberalism by freeing it from too close an association with *laissez-faire* economics, and from the association of morality with utility.[101] There was, in the view of many idealist liberals, a distinction to be made between the right and the wrong kind of socialism.[102] Henry Jones, whom Collingwood described as 'one of our most eminent philosophers',[103] frequently wrote and spoke upon this issue. The right type of socialism, or state intervention, was that which enhanced the individual's freedom by making possible activities, or modes of personal development, which free enterprise was ill equipped to provide, an example of which would be a world-wide postal system. He says, for example, that 'state organization empowers the individual. Under the new and more socialistic state, more men can say

---

[99] Id., *The Idea of History* (Oxford, Oxford University Press, 1973: first published posthumously 1946), 264, 126, and 122–3 respectively. The comments were written for *The Principles of History* on which Collingwood was working during his voyage to the East Indies in the early part of 1939.

[100] Id., *New Leviathan*, 33.46. Also see 12.95, 19.83–94, 25.33, 33.45, 33.76–99, 37.58.

[101] See e.g. B. Croce, *History of Europe in the Nineteenth Century*, trans. H. Furst (London, Allen and Unwin, 1934), 297, 298, 302, 306, 307, 309, 310, 313; de Ruggiero, *European Liberalism*, pp. 393–4; H. Jones, *Idealism as a Practical Creed* (Glasgow, Maclehase, 1909), 115, 218–19.

[102] Croce, *History of Europe*, pp. 298, 309; de Ruggiero, *European Liberalism*, pp. 32, 391–3; H. Jones, *The Working Faith of the Social Reformer* (London, Macmillan, 1910), 89–114; F. Anderson, *Liberalism and Socialism* (Adelaide, University of Adelaide Press, 1907), 1–10.   [103] Collingwood, *Speculum Mentis*, pp. 73–4.

mine of more things.'[104] True socialism enhances the individual's personality, extends the mind, and develops the moral self.[105] Even Croce, whose conception of liberalism is more narrowly confined than that of either Jones or de Ruggiero, acknowledged that ethical liberalism was compatible with many measures that may be classified by abstract theorists as socialist.[106] Indeed, Croce talks of a 'true' or 'genuine and effective' socialism, by which he means a socialism conceived in terms of the principle of liberty.[107] Fabian socialism appeared to Croce to be imbued with this liberal spirit.[108]

In essence, one might say that most of the idealist liberals were out of sympathy with the individualism, or libertarianism, of classical liberalism and were cautiously optimistic about extending the role of the state in accordance with a more collectivist view of its function in society. Indeed, among many of the idealist liberals there existed the belief that state intervention could, if appropriately and judiciously extended, enhance individualism, by which they meant the provision of greater opportunities for moral development, rather than the licensing of capricious action. Liberalism's involvement in government and administration had, according to de Ruggiero, made it far less hostile to the state. The hostility was further reduced in recognition that the state had become much more able to express synthetically the variety of interests articulated in society. Liberalism simply could not ignore the dire unintended social consequences of industrialization, and the inability of private entrepreneurial and charitable action to offer effective solutions. The state is the creation of those individuals over whom it presides, and in contributing positively to the increase in their well-being it strengthens itself.[109]

Given that Collingwood declared himself at one with the principles which de Ruggiero espoused we would expect the former to share the general idealist liberal tendency to acknowledge the value

---

[104] H. Jones, Dunkin Lectures on Sociology at Manchester College, Oxford, 25 Nov. 1904. The quotation is taken from R. V. Lennard's notes, Bodleian Library, Oxford, MS Top. Oxon. e. 417, fo. 8. Cf. 'The criterion of the action of the state is the effective freedom of its citizens', Jones, *Working Faith*, p. 113.

[105] H. Jones, 'True and False Socialism', a lecture given at Wollongong (NSW), Wednesday, 21 July 1908 and at Newcastle (NSW), Wednesday, 5 Aug. 1908. Reported in *The Illawarra Mercury* 24 July 1908; *The South Coast Times, 25 July 1908; and Newcastle Morning Herald*, 6 Aug. 1908.

[106] Croce, *Politics and Morals*, pp. 105–6.   [107] Id., *History of Europe*, p. 298.

[108] Ibid. 309.   [109] De Ruggiero, *European Liberalism*, p. 369.

of some forms of socialism, and to applaud the more extended role of the state in society as long as the principles of intervention were liberal in spirit. Collingwood actually says very little about forms of socialism which subscribe to the idea of change through democratic participation in politics, and tends on the whole to discuss only the form of socialism that other idealist liberals would call the wrong kind. Indeed, socialism was a vital force in the world, Collingwood believed, only in its Marxian form, by which he implies that it is an enemy to liberalism.[110] Nevertheless, he did seem to think that capitalism and socialism were necessarily co-existent 'tendencies in the economic organization of one single age', and that if either disappeared the other could not survive.[111] Marx made the mistake of seeing them in abstract opposition, rather than as complementary and mutually beneficial tendencies. Collingwood admits, for example, that the programme of socialism is in principle liberal,[112] but fears most of all, like all those liberals with whom he has an affinity, the potential in socialism for authoritarianism and class war. This fear did not make him averse to state intervention. The state, for Collingwood, is an empirical entity whose functions have been historically accumulated in response to the needs and demands of their communities. Although the state exists primarily to perform political works deemed necessary to the preservation of the body politic, its identity 'as a concrete historical institution' demands that 'it must to some extent interest itself in economic and moral questions',[113] which are an outgrowth of its primary task. For example, where business people manipulate the market for personal profit and to the detriment of the community there is a misuse of economic tools. In such circumstances where individuals' conscience has not prevented malpractice the state should intervene.[114] In addition, Collingwood believed that 'a heavy burden of taxation' should be reverted to in order to fulfil the functions of the state rather than attempt to increase the money supply and generate higher commodity prices. The consumer pays either way: explicitly

---

[110] Collingwood, 'Man Goes Mad', p. 25 (this book, chap. 14).

[111] Id., 'Lectures on Philosophy of History', p. 20.

[112] Id., 'Man Goes Mad', p. 25 (this book, chap. 14).

[113] Id., Lectures on 'Moral Philosophy, 1933', Collingwood MS, DEP 8, p. 99 (this book, chap. 6).

[114] Id., 'Money and Morals', Lecture to the Student Movement, London Branch, at 32 Russell Square, on 27 May 1919, and repeated at the Indian Students' Hostel at 21 Cromwell Road, June 1919, Collingwood MS, DEP 1, p. 10.

and consciously through taxation, or unawares through an increase in prices.[115] Whether the benefits of a sound currency are those we aspire to attain, or whether other aims incompatible with the idea should be pursued 'is a question of politics'.[116] Furthermore, the state should step in and prevent a person from having to accept a wage which in circumstances less degrading and without duress, he or she would not freely agree to accept.[117] It is also appropriate for the state to enter into slum clearance programmes and provide housing for the poor.[118] And, in the realm of morality, 'the state must have the power of suppressing seditious or obscene publications'.[119]

All of these areas of state intervention, and any others that may from time to time be appropriate, should be entered into with reference to a constant and consistent principle. All instances of state intervention should be perpetrated on political, and not moral or economic grounds. The state's purpose is to promote the political good of order, and it is charged with the task of entering into activities conducive to the promotion of this good.[120]

Political action, or regularian action, as we will see in the next section, is a universal category of action and cannot therefore be restricted to any particular sphere. The question of state intervention cannot therefore revolve around such issues as the intrusion of politics into areas of life which are clearly non-political: all areas of life have a political element, that is action according to rule, so the question becomes one of which political actions are best left to associations like the family, club, university, or business, and which are best dealt with by the state.[121] It is simply a mistake, as far as Collingwood is concerned, to think that 'governments can be kept

---

[115] Id., 'Economics as a Philosophical Science'; 'For a section of a comprehensive ethical treatise: or alternatively as a small book under the above title.' Collingwood MS, DEP 24, p. 23.

[116] Id., 'Economics as a Philosophical Science' (draft), Collingwood MS, DEP 24, p. 23.

[117] Id., 'Economics as a Philosophical Science', *International Journal of Ethics*, 35 (1925), 176.

[118] Id., 'Political Action', *Proceedings of the Aristotelian Society*, 29 (1928–9), 162 (this book, chap. 4).

[119] Id., 'Moral Philosophy Lectures [1929]', DEP 10, p. 107, (this book, chap. 5).

[120] Id., 'Stray Notes on Ethical Questions', 1928, Collingwood MS, DEP 6, p. 30; 'Political Action', p. 163. (this book, chap. 4); 'Lectures on Moral Philosophy, 1933', p. 100 (this book, chap 6).

[121] Id., notes on MacIver, *The Modern State*, comments on p. 149 of the book, Collingwood MS, DEP 20.

out of mischief by denying them certain powers'.[122] Limiting the extent of state intervention simply fails to address the fundamental problem of the nature of the relation between the state and its citizens. If the state is prevented from interfering in the private affairs of its citizens, society itself will invade that sphere, proving to be just as extensive in its effects, and just as difficult to justify as state intervention.

The problem can be overcome, Collingwood believes, by understanding the state as mind. The ruler and the subject are two different minds, but at the same time share an identity. In so far as they 'are united by common purposes, they have not two wills but one common or general will which is the measure of their partnership in a common life'.[123] The state is a mind

in the sense that any organized system of acts is itself an act, and that therefore any loyal member of a community wills the whole organization of that community, not merely the little part of it which is in an immediate sense his own business. The mind of the state is the mind of any member of the state, so far as he genuinely shares in its life.[124]

Political activity is, for Collingwood, self-government in so far as any person who devises a scheme for a society or community is an integral part of the community for which the scheme is devised. Those members of the community who are affected by the scheme choose by their own free will to comply with the new arrangements because no one can be compelled to acquiesce against their wills.

In summary, many of the idealist liberals discussed the question of state intervention, or interference, in terms of the conventional opposition of individualism, or libertarianism, and socialism, or collectivism. Each side of the opposition was, in their view, abstract in that each accentuated 'this or that element to the exclusion of the other'.[125] What was needed was a conception of the human condition

---

[122] Id., 'Economics as a Philosophical Science', p. 185 (this book, chap. 2).

[123] Id., 'Moral Philosophy Lectures [1929]', p. 127.   [124] Ibid. 128.

[125] De Ruggiero, *European Liberalism*, p. 157. Cf. Jones who captures the essence of what idealist liberals were trying to do when he says, 'Both the Individualist and the Socialist regard the State or civic community, and the individuals who constitute it, as more or less exclusive and independent of each other. The correction of errors comes from recognizing more fully that the State or the city and its citizens have only one life; so that each in repressing its opposite is destroying itself. In other words, the Individualist must be brought to see that his dependence on society is much more close than he deems and the Socialist that the welfare of society depends on providing for the individual the means for the most vigorous growth of an

which viewed the individual and society in an integral relation: each part nourished by the whole, and the whole being more than the sum of its parts. Socialism and collectivism were associated, favourably in the minds of the idealist liberals, with increased state involvement in enhancing the conditions of life for all the members of the community. As long as socialism accentuated, rather than retarded, the development of freedom, that is, as long as it was based on liberal principles, there could be no objection to it. Given Collingwood's attitude to state intervention and his professed admiration of de Ruggiero's political philosophy there is no reason to believe that Collingwood's own views would have been markedly at variance with those of the idealist liberals.

It was the authoritarian tendency in socialism, and its emphasis upon class interests and class war, that made socialism distinctly illiberal and therefore unattractive to the idealist liberals. Communism, appeared to them, to be the embodiment of the worst features of socialism. Croce, for example, argued that the ethics and politics of socialism were at base authoritarian in character.[126] Communism, under the guise of socialism, had proved to be, in Croce's view, a bitter enemy of liberalism which it mocked and derided as moralistic.[127] In its Russian manifestation Communism was simply 'a form of autocracy' less liberal than the Czarist autocracy that it succeeded.[128] Henry Jones's main criticism of the wrong type of socialism was that it appealed to one class and sought to promote its interest at the expense of other classes. He acknowledged that these other classes had in the past ruthlessly exploited the workers, but strongly contested that this iniquity was no justification to reverse the condition. He appears to have had no substantive difference with the Labour Party over policies, but could not reconcile himself to the Party's, as its very name implied, blatant class appeal. Jones objected to the idea of Labour members of Parliament because such representation perverted the spirit of democracy by rejecting the

---

independent personality—means which include, amongst other things, full rights of private property and full scope for private enterprise.' Jones, *Working Faith*, pp. 272–3. A. Simchoni discusses this aspect of British idealism in her 'British idealism: Its Political and Social Thought', *Bulletin of the Hegel Society of Great Britain*, 3 (1981), 16–31.

[126] Croce, *Politics and Morals*, p. 106.   [127] Id., *History of Europe*, p. 353.
[128] Ibid. 356.

notion of a common good in favour of the good of a class.[129] These sentiments were echoed by de Ruggiero who deplored the influence of socialist parties on modern political practices. The reduction of all conflicts to those of classes and their economic interests, de Ruggiero believed, threatened 'to frustrate every effort to express economic conflicts in the higher terms of politics and to discover an ordered and civilized *modus vivendi* for the classes'.[130] De Ruggiero attributed the neglect of political values by socialism to the doctrine of historical materialism which posits the economic as fundamental, of which all other factors, such as the religious, political, and moral, are reflections.

Collingwood's views on the wrong type of socialism are in harmony with those of his fellow idealist liberals. He contended that the economic interpretation of history destroys the reality of the fabric of law which constitutes the state. The concrete historical character of the law is replaced with abstract utilitarian ethics which leads to the substitution of economics for justice, 'and the declaration of a class war which is the explicit negation of the state'.[131] Class war is no less pernicious than the Prussian philosophy, which was the fundamental cause of the First World War, and no less catastrophic in its consequences.[132] The autocracy of the Prussian philosophy and the dictatorship of the proletariat are equally to be feared as the potential destroyers of the traditional liberal values of civilization. Class war and the dictatorship of the proletariat are at variance with the dialectic method of solving political problems which is the hallmark of liberal politics. Socialism, in its Marxist form, Collingwood argues, is an anachronism retaining outmoded and obsolete conceptions from a past age which are radically undialectical in character. The true heir to the dialectical method is

---

[129] H. Jones, 'The Corruption of the Citizenship of the Working Man', *Hibbert Journal*, 10 (1911–12), 155–78. Also see H. J. W. Hetherington, *The Life and Letters of Sir Henry Jones* (London, Hodder and Stoughton, 1924), 92; and Vincent and Plant, *Philosophy, Politics and Citizenship*, p. 155. In a letter to Sydney Webb, Jones explains that he has always been a champion of the working class, having emerged from its ranks himself, but he could not join the Labour party because of its emphasis upon economic class interest. He says: 'I love the working man too well to ask him to legislate primarily for himself.' Letter from Jones to Webb, 29 Mar. 1918. National Library of Wales, Thomas Jones Collection, Class u., vol. I, fo. 43.

[130] De Ruggiero, *European Liberalism*, p. 384.

[131] Collingwood, *Speculum Mentis*, p. 228.

[132] Id., 'Spiritual Basis of Reconstruction', pp. 14–16 (this book, chap. 17).

not, for all its protestations, Marxist socialism, but modern liberalism.[133] Indeed, liberal parliamentary democracy was, Collingwood
believed, 'an antiseptic against class war'.[134] It is certain that he
thought democracy less effective than it could be because its
principles had been undermined by a popular press and an unscrupulous government, but there is no reason to believe that Collingwood therefore thought that class war was a preferable political
method to the dialectic of liberalism. Shortly after writing *An
Autobiography* Collingwood reproves both Marx and Lenin for their
hostility to liberal ideals and refers to the idea that states have always
been the instruments of oppression in the class war as a 'monstrous
lie'.[135] When Collingwood's three laws of politics function within
the body politic effectively their implied and implicit purpose is to
dissipate the potential for class war.

The reason why Collingwood praises Marx in *An Autobiography* is
not because of any radical shift to the left in Collingwood's politics,
but because Marx refused to accept the false dichotomy, prevalent
among the realists whom Collingwood reviled, between theory and
practice. Marx was a philosopher who wanted to change the world,
and as far as Collingwood was concerned this was a laudatory
ambition for a philosopher to foster,[136] and one to which Collingwood himself had always subscribed in his attempt to bring about a
rapprochement between theory and practice.

I have contended that the principles of Collingwood's liberalism
varied little throughout his life and that the apparent shift to the left
which many of his friends and colleagues perceived during the late
1930s was illusory. What in fact occurred was an intensification of
the passion with which he applied those principles in the face of the
accentuated threat to liberalism. The fact that Collingwood supported the anti-Appeasement cause was tantamount to guilt by
association. To espouse such views was an invitation to be indiscrim-
inately labelled a Communist, which was compounded in Collingwood's case by his avowed admiration for Marx's fighting spirit.

III

It is now appropriate to consider Collingwood's conception of
philosophy and its implications for the relation between theory and

---

[133] Id., 'Man Goes Mad', pp. 21–6 (this book, chap. 14).
[134] Id., *Autobiography*, p. 159.
[135] Id., 'Fascism and Nazism', p. 171 (this book, chap. 15); *New Leviathan*, 12.95.
[136] Id., *Autobiography*, pp. 152–3.

practice in order to discern the place of politics in his understanding of action as a whole.

Philosophy, for Collingwood, can never begin in absolute ignorance, nor can it end in absolute certainty. Instead its aim is to make more intelligible, or to specify more adequately, something that is already understood.[137] When philosophers wish to make more intelligible the concept of action they must assume that they already know what it is. This knowledge is, however, confused and inadequate. As philosophers we begin by investigating very carefully how action is understood in everyday language, and go on by degrees to articulate a less confused and more intelligible account of the concept, rather than 'to acquire brand new ideas that have never been in our minds before'.[138]

Collingwood's conception of philosophical method was formulated in conjunction with his various writings on moral philosophy. The method was applied to experience as a whole in *Speculum Mentis*, in which each form of experience was deemed to have a correlative form of action appropriate to itself. Art was associated with play, religion with convention, science with utilitarian ethics, history with duty, and philosophy with absolute ethics.[139] The method also found expression in its application to art, economics, and politics[140] prior to being fully articulated and modified in *An Essay on Philosophical Method*. Collingwood viewed the *Essay*, not as a final theoretical statement, but as a prelude to its application in future work.[141] *The New Leviathan*, may be considered the fulfilment of this task in that the method is applied to the subject-matter in terms of which it was originally developed and conceived.

In formulating the method Collingwood wished to demonstrate that the philosophical concept had its own differentiae which served to disassociate it from the logic of the empirical and mathematical sciences. The traditional logic of classification distinguished a genus into co-ordinate species, all of which embodied the generic essence to an equal degree. Whatever its status in the natural sciences, and

---

[137] Id., *An Essay on Philosophical Method* (Oxford, Oxford University Press, 1933), 11, 100, 161, 163, 164, 168, 205.

[138] Id., 'Moral Philosophy Lectures [1929]', p. 10.

[139] Id., *Speculum Mentis*, pp. 102–7, 134–8, 169–76, 221–31, 304–5.

[140] See id., *Outlines of a Philosophy of Art* (Oxford, Oxford University Press, 1925), reprinted in A. Donagan (ed.), *Essays in the Philosophy of Art* (Bloomington, Indiana University Press, 1964); 'Economics as a Philosophical Science'; 'Political Action'.

[141] Id., letter to de Ruggiero, 7 Feb. 1934.

this is a question into which Collingwood did not enter, it was
simply inappropriate for the task of articulating the determinations
of a philosophical concept. Species of a philosophical concept are
not co-ordinate embodiments of the generic essence; instead the
species differ in both degree and kind. They differ in degree by
being more and less adequate specifications of the generic essence,
and differ in kind in that the specifications take different forms.[142]
If each species completely embodied the universal in equal degree
then the different specifications of the genus would have to be
explained in terms other than the universal itself. If we take religion,
for example, extraneous variables like economic, social, and political
conditions would have to be invoked to explain why, when each
specification embodies the generic essence perfectly, there are so
many different instances. The universal now becomes only a partial
element fused with the variety of extraneous elements in each
specification. The variety of religions, on the traditional logic of
classification, render the religious essence incapable of accounting
for its manifestations because it is only one of a diversity of elements
embodied in each religion. In order to overcome this problem the
apparently alien elements have to be understood not as extraneous
to the essence but as somehow derived from it. This entails
acknowledging that each and every religion is not an equal co-
ordinate manifestation of the generic essence, but that some are
more adequate embodiments than others.

The plurality of different religions now appear as differing in points of
religiousness: we can now say that the lowest religions hardly deserve the
name of religion at all, and that in the highest religions the ideal at which
all religion is aiming is now for the first time realized. And this is obviously
what we do say not only about religion but about most things. We speak of
certain men as being scarcely human: of political freedom or justice as
gradually evolved from rudimentary into more perfect forms: of the perfect
shaving soap or electric light bulb, and so forth, implying that others usurp
a name which strictly they do not deserve.[143]

In other words, each specification differs in degree and in kind from
all other species of the genus and can be delineated on a scale. This
in itself would not differentiate the philosophical concept from
similar scales formulated in the empirical and mathematical sciences.

[142] Id., 'Action', p. 39; *Essay on Philosophical Method* p. 57.
[143] Id., 'Action', p. 39.

Water, for example, can be represented on a scale in its solid, liquid, and gaseous forms in terms of the degree of heat that each embodies. The variable in the scientific scale is extrinsic to the generic essence. The generic essence of H_2O is equally embodied in all its forms, but the external variable of heat changes without affecting the essence.[144]

The philosophical concept is distinguished in that its 'variable is identical with the generic essence itself'.[145] This means that if there is a change in the degree to which the variable is present in a species there will be a correlative change in the degree to which the generic essence is present. The higher one ascends the scale the more adequately the form embodies the generic essence. Goodness, for example, is the generic essence of the philosophical concept of action and is present in a greater degree as you ascend the scale, and to a lesser degree as you descend it. If one takes the sequence of caprice, utility, right, and duty each is a more adequate embodiment of the generic essence. As the reasons one has for choosing between alternative actions become more rational, that is as the element of caprice is slowly eliminated, the more goodness the action embodies.

Unlike the methods of classification in the empirical and mathematical sciences in which the variable is measurable, in the philosophical concept the variable is not susceptible to quantification.[146] In the 1940 lectures on moral philosophy Collingwood makes a distinction between 'two kinds of muchness';[147] one which is merely a matter of degree and the other which is both a matter of degree and quantifiable. When there are differences of more and less they are only measurable if the 'muchness' is of the quantitative kind. He explicitly takes W. D. Ross to task for contending that 'if anything is greater than another thing, it must be greater by some definite amount'.[148] For goodness and pleasure to be measurable, as the Provost of Oriel claims, they must be commensurable. But, Collingwood argues, the Provost nowhere questions 'what the characteristic requisites of commensurability are'.[149]

Collingwood suggests that commensurability consists in the fact

---

[144] Id., *Essay on Philosophical Method*, pp. 59–60.   [145] Ibid. 60.

[146] Ibid. 70–1.

[147] Id., 'Goodness, Rightness, Utility: Lectures delivered in Hilary Term, 1940', Collingwood MS, DEP 9, p. 17.

[148] Ibid. 17, 25. W. D. Ross, *The Right and the Good* (Oxford, Oxford University Press, 1930), 143.   [149] Collingwood, 'Goodness, Rightness, Utility', p. 19.

that the things said to be commensurable are comprised of units all
of which must be measurable in terms of a common scale. Collingwood contends that the Provost does not show that pleasures have
this requisite, and in fact the Provost admits that they do not by
claiming that 'the difference between two pleasures is not a pleasure,
as the difference between two lengths is a length'. The latter is
commensurable and therefore measurable, but the former is incom-
mensurable because of the absence of common measurable units.[150]
We can only estimate or guess the extent of the differences between
degrees of goodness and of pleasure because there is never a mere
difference of degree, but always at the same time a difference in
kind. To suggest that our guesses are in fact measurements is simply
a misuse of language.[151]

The specifications of a philosophical concept are related not as
mutually exclusive species of a genus but as overlapping forms in a
linked hierarchy. Collingwood maintains that the division of philosophical concepts into mutually exclusive classes falls victim to the
disjunctive fallacy which in its positive form leads to the fallacy of
precarious margins, and in its negative form results in the 'fallacy of
identified coincidents'. In the positive form we find, for example,
that when classifying actions one very soon identifies those which
embody mixed motives and resist all attempts of exclusive classification. In other words there will be an overlap at the margins of
each specification. If one ignores this by identifying a margin which
has not succumbed to an overlap upon which we build a classificatory category on the assumption that it exhibits the pure essence of
the class, one falls victim to the fallacy of 'precarious margins'. This
is because to admit of an overlap relinquishes any grounds for
assuming it will not spread. If the disjunctive fallacy is applied
negatively, and one assumes that the overlap of classes is in principle
limitless, then we avoid the fallacy of precarious margins but
immediately embrace another, that is, the 'fallacy of identified
coincidents'. For instance, the observation that the general happiness is increased by the performance of one's duties leads the

---

[150] Ibid. 19–20. See Ross, *The Right and the Good*, p. 143. Collingwood goes on to
convict Ross of holding contradictory views on the question of the measurability of
goodness and pleasure. Cf. *Essay on Philosophical Method*, p. 78 n. 1, which implies
that Ross agrees with Collingwood's views on measurement.
[151] Collingwood, *Essay on Philosophical Method*, p. 71; Collingwood, 'Goodness,
Rightness, Utility', pp. 22–3.

utilitarian to conclude that there is no distinction to be made 'between the concept of duty and the concept of promoting happiness'.[152]

Collingwood contends that the relations among specifications of the philosophical concept are not disjunctive, but conjunctive; each class is therefore not exclusive but overlapping.[153] This means that the philosophical concept is transcendental'[154] in that instances of each of its species partake of and are instances of every other. The forms of experience illustrate the relation involved:

> They are activities each of which presupposes and includes within itself those that logically precede it; thus religion is inclusively art, science inclusively religion and therefore art, and so on. And on the other hand each is in a sense all that follows it; for instance, in possessing religion we already possess philosophy of a sort, but we possess it only in the form in which it is present in, and indeed constitutes, religion.[155]

Each of the forms is a fusion of differences in degree with differences in kind. The fusion is itself explained by the idea of overlapping classes. Differences of degree and differences of kind are both species of the genus difference and therefore overlap to form a difference that partakes of both.[156]

A concept may be distinguished into species whose relation to one another is that of opposition. For example, we may distinguish actions into good and bad. Alternatively the specifications of a concept may be related by distinction in classifying actions as just, generous, and courageous.[157] However, on the principle of overlapping classes the relation between specifications of a concept can never be pure opposition, nor pure distinction, but instead a relation which is at once a fusion of opposites and distincts. If the philosophical scale is constructed on the principle of opposition, zero and infinity mark the two ends. In other words if actions are distinguished into good and bad the extremes on the scale are absolute goodness and

---

[152] Id., *Essay on Philosophical Method*, p. 49.

[153] Id., 'Utility, Right, and Duty', Collingwood MS, DEP 6, p. 24. The essay is undated but parts of it are written on notepaper headed 'Admiralty War Staff, Intelligence Division' in which Collingwood served during the First World War. It is likely that some of it was written about 1920 with modifications added later in the same decade.

[154] Id., 'Moral Philosophy Lectures [1929]', pp. 6-7.

[155] Id., *Outlines of a Philosophy of Art*, ed. Donagan, pp. 144-5.

[156] Id., *Essay on Philosophical Method*, p. 74.   [157] Ibid. 64.

absolute badness. This is an impossible delimitation in a philosophical scale because of the fact that the variable is identical with the generic essence. Thus if goodness is the universal essence of action there can be no point on the scale at which goodness is absent without ceasing to be action. The scale cannot begin at zero, and must instead begin at unity in which there is a minimal embodiment of the generic essence, and in which there can be no absolute opposite to it. This is not to suggest that the scale is devoid of opposition and comprised entirely of distincts.[158] The minimum realization of goodness is distinct in relation to other realizations, and as the limiting case stands in opposition to the rest of the scale. Taken on its own each specification of goodness, however low, is good, but when compared with higher specifications it becomes thoroughly bad. The pursuit of pleasure as such is good, but when compared with the pursuit of duty it is thoroughly bad.[159] At every point on the scale there is a fusion of opposition and distinction.

Two questions remain to be answered. Why do different specifications of the philosophical concept come into being? And, why do they necessarily overlap?

Collingwood acknowledges his debt to Aristotle in contending that the generic essence is differentially present in its own species, but attempts to improve upon Aristotle by explaining why one specification should give way to its successor. The reason for the emergence of succeeding specifications is that each tries to be what it is not. There is a discrepancy between what a form of experience, or specification, is and what it strives to be. In trying to realize its true nature it becomes modified and transformed into something else. Action, Collingwood argues, is identical with self-knowledge. The driving force that impels each form of action to pass over into another is therefore the progression of self-knowledge: 'The mind comes to know itself in the course of its experience, and every stage in the growth of this knowledge is marked by the emergence of a new type of action.'[160]

The reason for the overlap in classes is related to the reason why new forms succeed the old. The lower of two forms on a scale possesses both general and specific goodness: general in so far as it exhibits the generic essence; specific in that it exhibits it in its own way. In comparison to the higher the lower loses both its general

[158] Ibid. 81.   [159] Ibid. 84.   [160] Id., 'Action', p. 42.

and specific goodness. In comparison to the lower the higher gains not only the general but also the specific goodness of the form it has succeeded. The higher of the two possesses both its own goodness and also that of the lower form. The lower purports to possess a kind of goodness which in fact it only approximates. The type of goodness which the lower purports to possess is only achieved in the higher which itself purports to possess a kind of goodness only achievable in the form beyond it. Thus the overlap of the lower by the higher is explained by the fact that the higher includes within itself the positive content of the lower. The lower overlaps the higher in that its positive content is reaffirmed in the higher. The lower, however, rejects what the higher adds to it. Collingwood contends that the overlap is not one of 'extension between classes, but an overlap of intention between concepts'.[161]

It is this conception of philosophy that provides the foundation upon which Collingwood's long-standing differences with philosophical realism are based. He was vehemently critical of the realists' ethical theories on two counts, both of which stem from his conception of philosophical method. First, Collingwood convicts G. E. Moore of constructing a theory of ethics upon the principles of classification derived from the empirical and mathematical sciences, and by implication of all the indiscretions associated with it. It is 'sheer illiteracy', Collingwood suggests, to search as Moore does for characteristics which belong to all right actions and no others, and for the common characteristics which belong to all good things, and belong to nothing but good things.[162] The method is one that H. A. Prichard, W. D. Ross, and E. F. Carritt habitually employ, but which is rendered totally futile by the principle of the overlap of classes.[163]

Secondly, Collingwood deplored the dichotomy that the realists had intruded between theory and practice. It was both pernicious and contemptible to suggest, as Cook Wilson had done, that knowledge makes no difference to the known object,[164] or to

---

[161] Id., *Essay on Philosophical Method*, p. 91.

[162] Id., 'Utility, Right and Duty', p. 24. See G. E. Moore, *Ethics* (London, Butterworth, 1912), 77–81, 133–8 for examples.

[163] Collingwood, 'Utility, Right, and Duty', p. 24.

[164] Id., *Autobiography*, pp. 44–5; Collingwood, of course, was extremely critical of Cook Wilson's refusal to publish. For Wilson's reasons for believing that the object is independent of thought and remains unaffected by it see R. Metz, *A Hundred Years of British Philosophy*, trans. J. W. Harvey, T. E. Jessop, and H. Sturt (New York, Macmillan, 1938), 522; and J. Passmore, *A Hundred Years of Philosophy* (Harmondsworth, Penguin, 1980: 2nd edn.), 244–6.

maintain, as Carritt does, that in ethics 'the theory is nothing but an attempt to describe the facts; and to tamper with them is to poison the wells of truth'.[165] On the principle of overlapping forms, practical reason and theoretical reason are species of the genus rational thinking and therefore must overlap, or fuse, to form a mode of reason that partakes of both.

In *An Autobiography* Collingwood contends that the achievement of a *rapprochement* between theory and practice had been one of the major preoccupations of his life.[166] The evidence certainly bears this out. In *Speculum Mentis* Collingwood argues that the purpose of our theories and thoughts about ourselves and the world to which we are related is not to advance further the life of pure contemplation, but to enable us more freely to disclose our natures in 'a vigorous practical life'.[167] If our theories are erroneous then they act upon the mind itself and affect our actions. The mind that harbours false conceptions attempts to live up to them.[168] The purpose of theory is not to provide people with ideals and codes of conduct,[169] but to identify and clarify misunderstandings in the hope that the way will be opened to the possible solution of problems.[170] Theory should inspire an optimism in the possibility of a satisfactory resolution of practical problems. What Collingwood was trying to impress upon his audience was that the subject's consciousness of his or her situation makes a difference not only to a particular aspect but to the whole.[171] This is why he sought to impress upon his students the imperative need for thinking clearly before acting because the more clearly one thinks the greater the possibility of acting better.[172] To be confused in one's thinking about what one is doing affects not just one's conception of the action, but the action itself.[173]

---

[165] E. F. Carritt, *The Theory of Morals* (London, Oxford University Press, 1952: first published 1928), 70. For Moore's contention that our awareness of an object makes no difference to it see 'The Refutation of Idealism', reprinted in G. E. Moore, *Philosophical Studies* (Oxford, Oxford University Press, 1922), 29. Collingwood criticizes Moore's statement in *New Leviathan*, 5.31–2.

[166] Collingwood, *Autobiography*, p. 147.   [167] Id., *Speculum Mentis*, p. 15.

[168] Ibid. 241, 250. Cf. *Autobiography*, p. 147.

[169] Id., *Essay on Philosophical Method*, p. 131; 'The Present Need of a Philosophy', *Philosophy* 9 (1934), 262 (this book, chap. 11).

[170] Id., 'Political Action', p. 158 (this book, chap. 4), and 'Present Need of a Philosophy', p. 263 (this book, chap. 11).   [171] Id., *Speculum Mentis*, p. 244.

[172] Id., 'Lectures on Moral Philosophy, 1933', pp. 127–30 (this book, chap. 12).

[173] Id., 'Action', p. 77. Cf. 'Will and reason, choice between alternatives and recognition that these are alternatives, are thus the same thing: they represent the

The way in which we overcome our misconceptions, or errors, in self-knowledge is through historical understanding. History, Collingwood contends, is the mind's self-knowledge of itself.[174] History enables the individual to identify those aspects of the past which have a bearing on the present, and dismiss those which do not. The process promotes self-knowledge and self-criticism which raises the mind to higher levels of rationality, which in turn equips the person better for action.[175] It is the practical problems and predicaments in our present circumstances that set the historical problems we choose to pursue. In Collingwood's view 'all genuine historians interest themselves in the past just so far as they find in it what they as practical men, regard as living issues'.[176] He goes as far as to suggest that ultimately the aim of history is not to know the past but to understand the present.[177] The situations in which we are constantly having to decide how to act are clarified and made more intelligible by recourse to history. In *An Autobiography* Collingwood maintains that 'the plane on which ultimately, all problems arise is the plane of "real" life: that to which they are referred for their solution is history'.[178] Collingwood came to believe that history held the key to overcoming 'the traditional distinction between theory and practice'.[179] History is able to do so because, unlike the realist conception of acquiring knowledge, it does not presuppose its object but enacts it. Thus mind and object, theory and practice, are unified.

For the complete discussion of this unity of theory and practice we have to turn to *The New Leviathan*. In *The New Leviathan* Collingwood introduces a principle which he appears to have acquired from his anthropological studies.[180] This he calls the Law of Primitive Survivals which states:

practical and the theoretical aspect, respectively, of something which could not exist unless it were both practical and theoretical. By willing, we become able to think; by thinking we become able to will.' Ibid. 78.

[174] Id., *Idea of History*, pp. 142, 174, 175, 202, 218.

[175] Id., 'Draft of Human Nature and Human History', Collingwood MS, DEP 12, p. 22.

[176] Id., 'Notes on HISTORIOGRAPHY written on a voyage to the East Indies', DEP 13, p. 12.

[177] Id., 'History as the Understanding of the Present', Collingwood MS, DEP 16, p. 1.

[178] Id., *Autobiography, p. 114.*   [179] Id., 'Notes on HISTORIOGRAPHY', p. 21.

[180] Id., [Fairy Tales] 'B II. Three Methods of Approach: Philological, Functional, Psychological', Collingwood MS, DEP 21, pp. 11–20. See E. B. Tylor, *Primitive Culture: Researches into the Development of Mythology, Philosophy, Religion, Art and Custom* (New York, Gordon Press, 1974: first published 1871), 6, 14.

When *A* is modified into *B* there survives in any example of *B*,
side by side with the function *B* which is the modified form of *A*,
an element of *A* in its primitive or unmodified state.[181]

The introduction of this law entails a qualification to the argument
of *An Essay on Philosophical Method*. Collingwood is no longer
suggesting that there is a complete overlap of forms because an
aspect of the lower form remains pure and uncontaminated by the
higher. This means that the fallacy of precarious margins has either
to be severely modified or abandoned altogether because the existence of a pure survival is an acknowledgement that there is a pure
margin which the overlap has not encompassed. The implication is
that his argument against the traditional logic of classification loses
its force, but that need not detain us here.

The principle of a linked hierarchy of overlapping forms and the
law of primitive survivals explain why theory and practice are so
integrally related. Practical reason and theoretical reason are both
forms of rational thinking and consequently must overlap. The law
of primitive survivals explains 'why theoretical reason always contains a primitive survival of practical reason'.[182] For Collingwood
rational thinking entails distinguishing between the self and the not
self. The distinction, he suggests, is primarily practical. The function of rational thinking is to ask and answer questions about why
we choose to do the acts we do.[183] People are reassured about the
knowledge they possess and about the things they do when they ask
the question 'Why?' When one intends to do something the question
why often arises. Practical reason supplies three distinct, but
overlapping answers that comprise a serially related linked hierarchy. We may satisfy ourselves that we wish to do a certain action
by suggesting that it is useful as a means to an end, or that it
conforms to a certain rule, or that it is our duty. These are the
utilitarian, regularian, and dutiful forms of practical reason that
provide the grounds of which their intended acts are consequents.[184]
When questions about the not self arise reason is primarily theoretical. Reason can never be purely theoretical because it is a modification of practical reason and by the law of primitive survivals includes

---

[181] Id., *New Leviathan*, 9.51.   [182] Ibid. 14.38.   [183] Ibid. 18.1.
[184] Ibid. 14.33.

within itself an unmodified element of that out of which it grew. If this were not the case it would be difficult to explain the tendency in theoretical thinking towards anthropomorphism. By thinking analogously we project the reasons we give for our own behaviour upon the movements of people and things identified as the not self.[185] Anthropomorphism is a primitive survival in theoretical reason which we can recognize and control, but can never eradicate. The practical problems which arise from relating the self to the not self are the concern of theoretical reason, and it is because of this that it can never divest itself of practical reason. The theoretical forms of thought are more completely dependent upon the practical than are the practical upon the theoretical.[186] For Collingwood real thinking 'always starts from practice and returns to practice; for it is based on "interest" in the thing thought about; that is, on a practical concern with it'.[187] In other words one's concern with things other than oneself is both theoretical and practical: one's practical attitude towards these things will differ in relation to one's attitude towards one's own actions, and the theoretical attitude will be influenced by and arise out of the practical. In consequence the three forms of practical reason have their corresponding forms of theoretical reason.

When means and ends are the terms in which we view our actions, asking the question why of a not self is really enquiring 'to what end?' The Greeks, Collingwood argued, viewed their own actions and their own relations to the world in utilitarian terms, and therefore understood their world in terms of the same principles. They believed that Nature had her ends and devised means to attain them. This theological view of nature, Collingwood contended, persisted into the Middle Ages. Similarly, if practical reason views the actions of the self, and the relation between the self and its world in terms of obedience to laws then the world itself becomes understood in similar terms. For the man of the Christian Middle Ages right action, that is, action according to rule, took precedence over utility, or action according to ends. The idea of the laws of nature and the obedience of natural occurrences to these laws is a reflection of practical reason in the theoretical reasoning of modern science. The highest form of practical reason is duty: it is the consciousness of oneself as an individual, not performing an action

---

[185] Ibid. 14.51–61.   [186] Ibid. 1.67.   [187] Ibid. 18.13.

as a means to an end, nor an action of a particular kind in conformity
with a law, but a unique act in response to a unique situation which
it is my duty to do and no one else's, and which I do, being the sort
of person I am, because it is the only action available to me. It is
perfectly rational action, or as rational as it can be, that is, almost
totally devoid of caprice.[188] Its theoretical counterpart is historical
thinking: 'For history is to duty what modern science is to right,
and what Greco-medieval science was to utility.'[189] Historical thinking is the exploration of a world of unique agents other than oneself
in situations uniquely their own, doing the only things they could
do in the circumstances. Both duty and history are 'the idea of
action as individual' and 'the consciousness of duty is thus identical
with the historical consciousness'.[190] On the principle that the
problems addressed by theoretical reason arise in, and the answers
to those problems are referred back to, practical reason, history has
an eminently practical dimension. In this respect history stands 'in
the closest possible relation to practical life'.[191]

It is with the levels of practical reason that the essays in this
volume are primarily concerned, and in particular with regularian
or right action. Before letting Collingwood speak for himself on
these matters a few clarificatory remarks are in order. Moral
philosophy for Collingwood is unequivocally related to the question
of freedom of the will, and the exercise of this freedom in the act of
choice. To choose something is to choose some good,[192] but not all
goods are of equal worth. Goodness is a matter of degree. There is
some goodness in everything and therefore nothing is absolutely
bad.[193] To distinguish things into good and bad is primarily the
province of practical reason. To identify something as good is to
declare that it meets a standard which has arisen out of the practical
purpose the agent entertains towards the object.[194] Things are good,
Collingwood suggests, not intrinsically, but because they are
chosen.[195] It is therefore imperative that any theory of action
addresses itself to the reasons for choice, because it is these reasons

[188] Id., 'Goodness, Rightness, Utility', pp. 70–1 (this book, chap. 10).
[189] Id., *New Leviathan*, 18.51.
[190] Id., 'Goodness, Rightness, Utility', p. 75 (this book, chap. 10).
[191] Id., *Autobiography*, p. 114.
[192] Id., 'Moral Philosophy Lectures, New MS 1932', Collingwood MS, DEP 7,
p. 104.   [193] Id., 'Goodness, Rightness, Utility', p. 14.
[194] Ibid. 16.   [195] Ibid. 30, 32, 35 (this book, chap. 3).

which constitute the grounds of conferring goodness. In other words moral philosophy asks the question why people choose the things they do, and the answers given for the choices comprise different forms or specifications of action which are fusions of differences in degree with differences in kind, and of opposites and distincts.

Choices of which the agent is conscious can be distinguished into two types. First, there is mere choice, or capriciousness,[196] where the agent is conscious of freely choosing between alternatives, but unware of the reason why one alternative is chosen and not another. Capricious choice is an act of free will and demonstrates that the agent has liberated himself, or herself, from subjugation to the whims of slavish desire. Caprice is choice, but it confers a low degree of goodness. There is, however, a second type of choice 'in which the agent is conscious of having reasons for choosing'.[197] It is in this context that the distinction between theoretical and practical reason arises. Collingwood spent over twenty years constantly refining his theory of action and his various attempts at specifying the different forms of rational action, or the determinations of practical reason, can be distinguished into two broad categories: those in which duty and right are identified as constituents of the same form of action, and those in which they are specified as distinct but overlapping forms of action differing in both degree and kind. So far I have emphasized only the latter category. It is now appropriate to give a brief indication of both specifications.

It is commonly argued that goodness and duty stand in the closest relation to each other. Nothing can be our duty unless the action is good in itself, brings about good consequences, or some combination of both.[198] Furthermore, and in different ways, it was usual to associate what is right with what is morally obligatory. One's duty is to do what is right, and what is right is dictated by the rules of morality. Richard Price, for example, contended that '*obligation* to action, and *rightness* of action, are plainly coincident and identical', while Carritt maintained, 'my duty is to be moral, and that is to try to do what is right'.[199] For Collingwood, all of the forms of reasoned

---

[196] Ibid. 36; *New Leviathan*, 13.12.

[197] Id., 'Goodness, Rightness, Utility', p. 38, and *New Leviathan*, 14.1–11.

[198] See H. A. Prichard, *Moral Obligation: Essays and Lectures* (Oxford, Oxford University Press, 1947), 142.

[199] R. Price, *Review of the Principal Questions in Morals*, ed. D. D. Raphael (Oxford, Oxford University Press, 1948) 105; Carritt, *Theory of Morals* p. 94. For

action are ethical in so far as they embody a degree of goodness. In the 1923 lectures on moral philosophy, and in *Speculum Mentis*, Collingwood associates play and convention with capricious choice, and articulates three forms of reasoned choice. The first is utilitarian ethics, or economic action. Philosophically rather than empirically conceived, economic action is not a separate sphere of life, but a universal characteristic of all action. All action is purposive in so far as it can be conceived in terms of means and ends.[200] Utilitarianism abstracts from concrete actions only their purposiveness and calls it utility. Two elements are evident in economic action, immediate and mediate acts. We do something that we do not want to do, the immediate act, as a means to bring about something that we do, the mediate act or end.[201] Economic action is rational to a degree but still retains a significant element of capriciousness. We are aware of our reason for choosing an act, that is because it is a means to an end, but we are unable to give reasons for the choice of the means from among a variety of alternatives all of which would attain the same result, nor are we able to give reasons for the end chosen. Answers to these questions can only be achieved at a higher level of rational consciousness.[202] In characterizing the succeeding determination of action which Collingwood calls duty, or concrete ethics, he allies himself with those moral theorists, including the realists, who posit an identity between right and duty. He says, for example: 'an act is an act of duty, and its goodness is rightness'.[203] Similarly, politics, the moral science associated with right, or regularian action (action according to rule), is inextricably related to duty: 'the right to rule is also a duty to rule; and hence politics is so far a branch of ethics that duty is an indispensable part of the concept of the rule'.[204]

further identifications of right and duty see G. E. Moore, *Principia Ethica* (Cambridge, Cambridge University Press, 1903), 89; Moore, *Philosophical Studies*, p. 320; H. W. B. Joseph, *Some Problems in Ethics* (Oxford, Oxford University Press, 1931), 59, 96–8; H. Rashdall, *Theory of Good and Evil* (Oxford, Oxford University Press, 1907) i. 135, 138, and *Ethics* (London, Jack, no date), 12–14. Collingwood accuses both Moore and Rashdall of further reducing duty and rightness to utility in that their theories suggest that 'to call an act good is at bottom to call it useful'. Collingwood, 'Moral Philosophy Lectures [1929]', p. 97.

[200] Collingwood, *Speculum Mentis*, p. 171.
[201] Id., 'Economics as a Philosophical Science', pp. 167–8 (this book, chap. 2).
[202] Id., 'Action', 1923 conclusion, p. 66.   [203] Ibid., 1927 conclusion, p. 85.
[204] Ibid., 1923 conclusion, replacement section on law, p. 70.

Whereas utility in this sequence of forms of action is associated with scientific thought, duty or concrete ethics is related to historical thought. The question which remained unanswered in the economic form of action, namely 'why do we choose ends?' finds its answer in duty: the ends are chosen because they are good in themselves.[205] Historical thinking is the world of fact which partially succeeds in unifying the universal and the particular. In duty the means and end of economic action coincide, 'the means becomes an end in itself and the means becomes the means to itself'.[206] However within this form of action an opposition exists which can only be resolved in the succeeding form. Individual conscience intuitively grasps what its duty is, but conscience is not able to give an account of itself, nor explain the multitude of differing and conflicting intuitive pronouncements of duty emanating from other individuals. On the other hand and in opposition to individual conscience stands an objective moral order embodied in the system of law which stands over and directs the conduct of the subjective will. Conscience is opposed by authority and individual freedom is restricted by a force external to itself. In this respect duty is not wholly concrete and still retains an element of the abstractness of scientific thought, or the abstract ethics of economic action.[207]

Beyond the realm of right and duty Collingwood posited an immanent form of action which he called Absolute Ethics, Absolute Will, or Absolute Action. Individual conscience and external authority are reconciled when the ruled and the ruler become united in a society where everyone can be relied upon to do what is right without command.[208] This does not mean that there is an absence of command, merely that there is a recognition and acknowledgement that both our commands and those of others are responsible because of an appreciation that the rationality one claims to possess oneself is also possessed by others.[209] Here we have passed into the realm of philosophy in which the agent embodies and is identified with absolute mind, that is, conscious of the fact that the mind creates its own objects overcoming at last the false dichotomy between will and intellect, or practice and theory. In the 'pure act of self-creation' the act and self-knowledge, will and intellect, are identical.[210]

---

[205] Ibid., 1923 conclusion, p. 64.   [206] Id., *Speculum Mentis*, p. 222.
[207] Ibid. 222–4; 'Action', 1927 conclusion, p. 85, and original conclusion, p. 69.
[208] Id., *Speculum Mentis*, p. 230.   [209] Id., 'Action', 1927 conclusion, p. 87.
[210] Id., *Speculum Mentis*, p. 305.

The determination of the generic essence of action in terms of the forms of utility, duty, and absolute ethics, in which duty and right were inseparable, was a formulation with which Collingwood became increasingly dissatisfied. He acknowledged the importance of Kant's contribution to enhancing our understanding of duty, by arguing that Kant, in insisting upon the autonomy of the will, had demonstrated that only the will is good without qualification, whereas the goodness of other things or forms of action is relative to something else; that is, they are useful, or good in relation to an end, and right in being good in relation to a rule.[211] Furthermore, Kant's principle that ought implies can emphasizes the very important point that if something is said to be someone's duty then it must be possible for that person to perform the act.[212] In these principles, Collingwood suggests, we have the germs of a non-regularian conception of duty. Unfortunately, Kant's 'theory of duty is a regularian theory' and the germs of the non-regularian conception are 'embedded in a hostile context'.[213]

The realists Carritt and Prichard both associate right with duty and attempt to draw certain consequences out of the principle that ought implies can. They contend that we can never be sure of our ability to perform an action and therefore it can never be our duty to do that particular act. Instead our duty must be only to try to do the act. For Carritt, 'my duty is to be moral, and that is to try to do what is right'.[214] In Prichard's terms we do not have an obligation to *do* something, but instead 'that of settling or exerting ourselves to do something, i.e. to bring something about'.[215] Collingwood is critical of Carritt's and Prichard's answer to the question 'what kind of things do we have a duty to do?' He suggests that what they are asserting is that 'it is never a duty to will an act simply, but always to will to will that act'.[216] Their way of stating the idea of the will to will, Collingwood argues, is open to the objection that the will has

[211] Id., 'Moral Philosophy Lectures [1932]', p. 105. See also id., 'Lectures on Moral Philosophy for M-T 1921', DEP 4, p. 108.

[212] Id., 'Goodness, Rightness, Utility', pp. 64–5; *New Leviathan*, 17.6–62.

[213] Id., 'Goodness, Rightness, Utility', pp. 64, 66; cf. *New Leviathan*, 17.62.

[214] Carritt, *Theory of Morals*, p. 94.

[215] H. A. Prichard, 'Duty and Ignorance of Fact', Annual Philosophy Lecture, Henriette Hertz Trust, British Academy, 1932: reprinted in *Moral Obligation*, pp. 34–5.

[216] Collingwood, 'Lectures on Moral Philosophy, 1933', p. 118. It is a 'distinction between willing and willing to will', ibid.

become an abstraction divorced from any content. To avoid this abstraction it is necessary to insist that 'the will to will is the will to will some definite act, so that without some definite act there is no will to will'.[217] In other words it is simply not good enough to try to do our duty because the will is only known by its achievements; we must will a definite act.

Even if Prichard's point is conceded his moral theory still fails, Collingwood contends, by falling into a second difficulty which arises out of W. D. Ross's observation that two actions may both be right,[218] but not compossible. If it is my duty to do what is right, whichever of the two compossible actions I choose to perform will involve a failure to perform my duty in relation to the other. If right and duty are identified the principle of 'ought implies can' does not hold because both acts ought to be done; both are one's duty in that it is right to do both, but it is only possible to do one or the other.[219] Collingwood was particularly impressed by Ross's recognition that right and duty are not synonomous terms, but disappointed that Ross did not make more out of it.[220] After acknowledging that '"right" has a somewhat wider possible application than "something that ought to be done"', Ross goes on to ignore the distinction for matters of convenience.[221]

Prior to the appearance of Ross's *The Right and the Good* Collingwood published a brief indication that he no longer believed it fruitful to associate morals and politics as he had done, for example, in *Religion and Philosophy* and *Speculum Mentis*. Both Machiavelli and Croce had, of course, clearly distinguished the realms of morals and politics, but had done so by resolving politics into utilitarian, or economic, action. In his article 'Political Action' Collingwood at once disassociates himself from those who identified economics with politics, and from those who conflated politics and duty. He argued that:

---

[217] Ibid. 120.

[218] Id., 'Goodness, Rightness, Utility', p. 65. See Ross, *The Right and the Good*, pp. 3–4.

[219] Collingwood, 'Goodness, Rightness, Utility', p. 65. Prichard had told Collingwood in 1933 that what is right is not always one's duty. Collingwood's criticism may therefore be unfair. See Collingwood's letter to H. A. Prichard dated 12 Feb. 1933, Correspondence of H. A. Prichard.

[220] Collingwood, 'Moral Philosophy Lectures [1932]', p. 93; 'Lectures on Moral Philosophy, 1933', pp. 107, 117; 'Goodness, Rightness, Utility', pp. 47–8.

[221] Ross, *The Right and the Good*, p. 3.

Political action, as such, is not moral action. A society makes a law not because it is its duty, but for some other reason—a political reason. Again, political action, as such, is not economic action. A society makes a law not because it will thereby become more wealthy, but because it will thereby achieve a good of another kind—a political kind.[222]

Collingwood had tried to articulate these distinctions earlier but declared himself dissatisfied with the result.[223] However, the 1929 lectures on moral philosophy and the subsequent revised courses of 1932, 1933, and 1940 all assumed the distinctions which found their fullest published expression in *The New Leviathan*; namely, that there are 'three moral sciences, economics, politics, and ethics . . ., which deal respectively with utility, rightness, and duty'.[224]

In this new sequence of forms of action duty is not merely separated from right. Absolute ethics is not in fact abandoned but is now part of what Collingwood means by duty. Whereas absolute ethics had been associated with philosophy in *Speculum Mentis*, duty which includes within itself absolute ethics becomes the practical counterpart of history, the highest form of theoretical reason in *The New Leviathan*. There appears to be no evidence to support Lionel Rubinoff's contention that this historical ethic, as Collingwood conceived it, would be superseded by a philosophical ethic as it had been in *Speculum Mentis*.[225] Rubinoff has allowed himself to be deceived by the fact that duty in both *Speculum Mentis* and *The New Leviathan* is associated with history. The supersession of history by philosophy in the former is not sufficient evidence for believing that Collingwood envisaged a similar pattern when he wrote the latter. The inclusion of Absolute Ethics in his new formulation of duty suggests that the former's theoretical form of reason is also absorbed by that of the latter. In the 1929 lectures on moral philosophy, for example, Collingwood makes the synonymity explicit by discussing the highest level of practical reason under the heading of 'Duty, or Absolute Ethics'.[226] Further, in *Speculum Mentis* Collingwood believed that the false dichotomy between the mind and its objects is overcome in philosophical thinking. By the time he wrote *The*

[222] Collingwood, 'Political Action', p. 159 (this book, chap. 4).
[223] Id., 'Utility, Right and Duty', p. 50.
[224] Id., 'Lectures on Moral Philosophy, 1933', p. 99 (this book, chap. 6).
[225] L. Rubinoff, *Collingwood and the Reform of Metaphysics: A Study in the Philosophy of Mind* (Toronto, University of Toronto Press, 1970), 171–2.
[226] Collingwood, 'Moral Philosophy Lectures [1929]', p. 129.

New Leviathan it was history which facilitated the reconciliation.[227] If in Collingwood's second series of forms of action, that is, utility, rightness, and duty, duty is 'as rational as activity can be', and 'triumphs over every form of its own opposite passivity' there would appear to be no room for a philosophic ethic.[228] Collingwood does not suggest that there will be no further development of the European mind, but he similarly did not speculate about which direction the development might take simply because such predictions are not possible.[229]

The details of the three kinds of reason for acting, or the three forms of goodness, need not be rehearsed here because sufficient indication of their content appears in the articles and extracts printed in this volume. It suffices to say that, empirically conceived, each of the three moral sciences, economics, politics, and ethics, artificially abstract limited areas of action within which the stated activity is believed to be more or less confined. Philosophically understood each form of action is a characteristic of every concrete act.[230] Thus, for example, the making of rules and conformity to rule cannot be confined to the state, but is universal in its manifestations, and the same is true of both utility and duty.

Utility and rightness retain within themselves differing degrees of capriciousness, and traces of caprice cannot be totally removed from duty. Utility is abstract because any one of a host of useful acts may lead to the same end. Positively it emphasizes the purposiveness of every act and the interrelatedness of means and ends. Negatively this form of action fails to explain the choice of end and the choice of means, from among many plans, to achieve the end. The performance of the means is merely the fulfilment of the condition to attain an end. Any other characteristics which the action may have are 'irrelevant to its utility'.[231] An act is right in that it conforms to a rule, but any other feature the action may possess is irrelevant to its rightness. Rightness cannot explain why I conceive myself to be the type of person I am, and as a consequence why I take rules of a certain kind to be applicable to the situations in

[227] Id., *Speculum Mentis*, pp. 249–50; 'Notes on HISTORIOGRAPHY', p. 21: 'In history the object is enacted and is therefore not an object at all.'
[228] Id., 'Moral Philosophy Lectures [1929]', p. 129.
[229] Id., *New Leviathan*, 9.43.
[230] Id., 'Moral Philosophy Lectures [1929]', p. 142.
[231] Id., 'Lectures on Moral Philosophy, 1933', p. 106. Cf. *New Leviathan*, 15.71–15.8.

which I find myself. No situation is ever so clear-cut and well defined that only one rule is applicable to it. A rule, Collingwood argues, is a 'disjunctive' imperative because of the many forms into which the rule may become differentiated and because each may be obeyed by performing a number of alternative actions.[232] Rightness leaves unexplained the choice of which rule to obey, and asserts that an act is right in that it is 'an act of this kind, not by being this act'.[233]

Utilitarian action is useful, or good, because it relates to an end. Right action is good because of its relation to the rule of which it is an instance. But both forms of action are, however, in differing, degrees, incompletely determinate. Duty, or the good will, 'is good in itself and apart from any relations whatever'.[234] A right act is one that any person can perform without detracting from its rightness. A duty, however, is never general, it is always someone's and no one else's; 'consequently the spirit in which the act is done, though it does not matter to its rightness, does matter from the point of view of duty'.[235] Duty is completely determinate and allows of no alternatives. Thus the element of caprice has been eliminated as far as it is possible.[236] My duty on any given occasion can only be discharged by me and involves the whole of my character in so far as it is the only act appropriate for me to do in the circumstances.[237]

In any given body politic the activity of ruling can be understood in terms of capricious action and all three levels of practical reason. These are what Collingwood calls the forms of political action.

Political action, Collingwood suggests, is the joint will of a society exercised by the ruling element in a body politic. In relation to the ruling element that will is exercised immanently in that it is self-rule. The same will appears as force when exercised transeuntly, that is, when the ruling class imposes its will upon the ruled. Without transeunt and immanent rule, which implies a ruled and ruling class, there can be no body politic. It is not a prerequisite of political life that the joint will should be rational. A decree in traditional political terminology is a command which is executive

---

[232] Id., 'Goodness, Rightness, Utility', p. 66; *New Leviathan*, 16.63.
[233] Id., 'Lectures on Moral Philosophy, 1933', p. 107; *New Leviathan*, 16.62.
[234] Id., 'Moral Philosophy Lectures [1932]', p. 105.
[235] Id., 'Lectures on Moral Philosophy, 1933', p. 109.
[236] Id., *New Leviathan*, 17.51, 17.55.
[237] Id., 'Lectures on Moral Philosophy, 1933', pp. 108, 114; *New Leviathan*, 17.8.

and not legislative in character and which is issued by those who rule to those over whom they exercise rule. Collingwood contends that: 'The decree is the simplest form of political action because it represents the simplest form of will, namely caprice, transposed into the key of politics.'[238]

Political action conceived in terms of utility entails distinguishing means and ends. The ends are determined by the ruling class and imposed upon the ruled class over which it presides. The means and the end are logically related in that the enactment of the former brings about the latter. Given this relation only the means need be divulged to the ruled and the end can remain concealed.[239] Utilitarian political action is policy, but the principles which guide it make both the choice of ends and means capricious. With regard to the external policy of the state differences between bodies politic are understood by rulers in terms of a conflict of interests. Understood in terms of right, political action entails translating the universal characteristic of rule-making into the formulation and enactment of legislation, or law. The element of caprice is evident in that different laws define different ways of life. One law defines one way of life the body politic should follow, and an alternative law defines another way of life. Both are capable of being conformed to by the body politic. The choice of which alternative to follow is a matter of caprice. Differences between bodies politic in external relations are understood by the rulers in terms of a conflict of rights.[240] When political action is understood in terms of duty the elements of caprice associated with utility and right are overcome. The way of life towards which the ruling class directs the body politic will be the only way of life that presents itself given the traditional values, character, and history of the people.[241] All alternatives would be repudiated and the duty of the rulers would be completely determinate. In external matters differences between bodies politic would be viewed by the rulers as a conflict of duties.

---

[238] Id., *New Leviathan*, 28.3. [239] Ibid. 28.52.
[240] Ibid. 29.58. [241] Ibid. 28.89.

PART ONE:

Political Activity and the Forms of Practical Reason

Editor's Comments

There has been little need for editorial intrusion into Collingwood's text, except to standardize spellings and identify sources for which the text gives no, or inadequate, references. Collingwood, although an accomplished and inspiring lecturer, delivered both lectures and papers from fully prepared scripts. Where footnotes have been added to the published and previously unpublished papers they are distinguished from Collingwood's own by appearing in square brackets.

The aim of Part One of this book is to allow Collingwood to place political activity in the context of action as a whole, establish its relations with the other forms of rational activity whose choices are explicable in terms of reasons, and with capricious action which is consciously free choice, but for which no reason can be given. The essays demonstrate clearly how each form of action differs in kind and degree from the others, and also how each overlaps, or implies, the others: each is distinct but none is separate and autonomous.

The essays were written at different times and for different purposes and therefore differ in the sequence of kinds of action they identify and specify. 'Economics as a Philosophical Science' was published in 1925, one year after *Speculum Mentis* and at the same time Collingwood delivered the moral philosophy lectures in which he discusses utilitarian ethics, duty, and absolute duty as the three forms of ethical action. 'Economics as a Philosophical Science' distinguishes economic activity from politics and morals, but does not attempt to establish distinct differentiae for moral and political action; the two inhabit the same sphere of activity without exhibiting a difference in degree or kind. Collingwood does not go on to discuss absolute ethics, but allows for the fact that there may be other specifications which are beyond the scope of his enquiry. In 'Goodness, Caprice, and Utility', an extract from the moral philosophy lectures delivered in 1940, Collingwood is assuming the sequence of caprice, utility, rightness, and duty which are the determinations of actions explicated in *The New Leviathan*, and which he formulated with greater and greater clarity in the lectures of 1929, 1932, and 1933. One theme which emerges in this extract is the question of the proper sphere of psychological explanation.

Collingwood had addressed himself to this question on many previous occasions. In *The Idea of History*, for example, he had made a clear distinction between feelings and thoughts, assigning the former to the province of psychology. In *An Essay on Metaphysics*, Collingwood castigates psychology for overstepping the bounds of its legitimate subject-matter and purporting to be a science of thought. As a so-called, or pseudo, science of thought it fails to provide a criterion by which acts of thought can be judged successes or failures, and attempts to usurp those sciences of the mind, like economics, politics, and ethics, that are criteriological. On the principle of overlapping classes the distinction between feeling and thought cannot be as rigid as is implied in *The Idea of History* and *An Essay on Metaphysics*, and indeed is not sustained in the lectures on moral philosophy, nor *The New Leviathan*; here feelings and thoughts overlap at each level of consciousness. In the extract from the 1940 lectures it is interesting that Collingwood does not assign the whole of capricious action to psychology, for elements of caprice persist in the higher levels of rational consciousness. Psychology has its uses in explaining to us, in terms of psychological laws, the emotional impulses which inspire 'morbid' actions, that is, those capricious actions which serve to impede, or frustrate, our attempts to live a certain type of life.

'Political Action' is principally concerned with establishing the differentiae of this distinct form of action, while at the same time deploring attempts to reduce it to the economic, and avoiding its subsumption under the category of morality. The theme of distinguishing between the empirical and philosophical sciences introduced in 'Economics as a Philosophical Science' is further explored in this essay and in 'Politics [1929]', but finds its clearest articulation in 'Politics [1933]'. All three essays on the political form of action explore the idea of rightness and its relation to politics by looking at different aspects of understanding conduct in terms of rules and obedience to rules.

Of the two essays on punishment which follow, the first, published in 1916, clearly identifies punishment and morality by suggesting that the former is the expression of the moral condemnation of the criminal. Similarly, desert as a criterion of who should be punished and by how much is described as a moral criterion. Collingwood disassociates punishment from economic action by making a sustained attack on the deterrence theory. His own understanding of punishment combines the retributive and reformative theories. In justification of punishment Collingwood gives the retributivist answer that it is the necessary expression of moral condemnation. In relation to the questions of who should be punished and by how much, Collingwood offers the retributivist's criterion of desert. But in terms of the purpose of punishment, or the end it should achieve, he offers the reformative answer; that is, its purpose is the moral regeneration of the criminal and of society. The second of the two essays on punishment, written in 1929, deals with the question of punishment in terms of an overlapping scale of forms in which the three forms of rational action, utility, right, and duty, are associated with the three principle theories of punishment; deterrence, retribution, and reform. The essence of punishment is understood in terms of retribution and political activity, but because of the principle of overlapping forms, while not being the essence of punishment, deterrence and reform may be its logical properties.

'Monks and Morals' and 'Duty' constitute two quite different discussions of the highest form of practical reason, yet both were written at about the same time. In the first, Collingwood is concerned to establish that ways of life, in addition to being judged by the criterion of social utility, may be judged in terms of intrinsic worth. The fact that he does not mention that ways of life may be judged in terms of their rightness is of little significance because nothing in what Collingwood says here excludes this possiblity. The second discussion of duty admirably illustrates the overlap of classes by showing that degrees of obligation are appropriate to both utility and right, and not confined simply to duty as a form of action. This extract from the 1940 lectures provides the best available summary of the indeterminateness of utility and right, and the complete determinateness of duty. It also shows much better than the discussion in *The New Leviathan* the relation between the highest level of practical reason, namely duty, and the highest form of theoretical reason, namely history. Neither history nor duty, Collingwood suggests, has developed to its full potential, that is, the idea of action as completely individual and completely determinate from which all elements of caprice have vanished.

2
Economics as a Philosophical Science

The science of economics enjoys today so vast a prestige, especially as compared with the philosophical sciences, that an attempt to treat economics philosophically might seem almost an insult. But the writer of this paper intends no such disrespect to existing economics as would be implied in a philosophical treatment of the empirical laws which economists reach by induction from the study of facts. What is intended is rather to throw light on some of the fundamental conceptions which economists do not so much derive inductively from facts as presuppose in their attitude toward the facts. These fundamental conceptions, such as value, wealth, and the like, are used by all economists, but seldom, if ever, satisfactorily defined. Yet in spite of this, people seem to know what they mean by them, much as people seem to know what they mean by space or time or identity or causation, though to define these is notoriously difficult.

These fundamental or presupposed conceptions of economic science are the subject of this paper. The thesis here advanced is that these conceptions are various aspects of, or various attempts to describe, a certain form of action which, for the sake of a provisional name, we shall call economic action. The conceptions of value, wealth, and so forth are not ultimate inexplicables; they can be understood, but only by resolving them into the conception of economic action. This resolution is a task for philosophy. Philosophical thought is that which conceives its object as activity; empirical thought is that which conceives its object as substance or thing. Economics, then, is an empirical science if it is conceived as the study of a thing called wealth; philosophical, if it is conceived as the study of economic action. But it is not enough, to make a science philosophical, that it should *call* its object action; it must really *think* of it as action. If a certain type of thought calls its object activity but nevertheless deals with it as if it were a substance or thing, it is an example of empirical, and not of philosophical, thought, and this

Reprinted from *International Journal of Ethics*, 35 (1925), 162–85, with the kind permission of the The University of Chicago Press.

is what happens in the case of psychology. Empirical economics, as belonging to the empirical study of human conduct, may be regarded as a branch of psychology: and just as there is a philosophical science of knowledge, distinct from the psychological science of knowledge, so there is a philosophical science of conduct (ethics) distinct from the psychological science of conduct. It has been recognized by philosophers in all ages[1] that the special type of action which economists study empirically is a type whose individuality is, from the philosophical point of view, both genuine and important; and therefore no apology is needed for raising once more the problem of philosophical interpretation of economic concepts.

The thesis here to be advanced, then, is that there is a special type of action, which we ordinarily distinguish by such epithets as expedient, useful, profitable, and the like; that this utilitarian or economic type of action is the fundamental fact with which all economic science is concerned; that a pure or philosophical economics would consist of the analysis of this special type of action and its implications; and finally, that the ultimate or fundamental problems of economics are soluble only by abandoning any attempt to solve them empirically or inductively, and dealing with them according to their true nature, as philosophical problems to be approached by philosophical methods.

Economic action is distinguished on the one hand from moral action, on the other from certain types of action whose characteristic quality is their immediacy. As opposed to moral action, economic action is not affected by the idea of duty or moral obligation. That

---

[1] To quote a few outstanding examples: Plato, followed by Aristotle, distinguishes between moral and economic action, and attaches great importance to the distinction; the state, for both these philosophers, is a moral activity which has its roots in, or grows out of, economic activities, and the essence of moral action is that it should at once control and safeguard economic action. Machiavelli and Hobbes did their best work in the analysis of economic activity and the insistence upon its reality and necessity. Kant sketches a philosophy of economic action when he lays down the doctrine of a hypothetical imperative, distinct from the categorical imperative of morality; Hegel works this hint out, with help from Plato and Aristotle, into a complete theory of economic society as a distinguishable 'moment' within political society. And Croce, while sweeping away most of the Hegelian distinctions as 'merely empirical', insists upon the philosophical validity of the concept of 'economicity', and regards the useful as a fundamental philosophical notion on a level with the beautiful, the true, and the good. It may perhaps be added that modern economics is, historically speaking, an offshoot of ethics, and was first taught from academic chairs of moral philosophy; hence the economist need not hesitate to obey the oracle *antiquam exquisite matrem*.

there are actions into which this idea does not enter is generally admitted. To invest money in a stock chosen for its promise of good dividends, or to obey a law or other command for fear of punishment may be right or wrong when judged by moral standards; but such actions can be done, and frequently are done, without the agent's raising any moral issue whatever. It may be argued by a rigid moralist that no one ought to invest money in a concern without asking whether he ought do so; which would involve inquiring into its aims and methods of doing business, and satisfying himself that they are morally praiseworthy; but no one will suggest that this inquiry is one which investors always, or as such, undertake. Not only, however, are some acts done without the agent's asking himself whether they are morally right, but a non-moral element is distinguishable within actions where this question has been raised. Thus, a man may invest money under a sense of obligation to provide for his children, and in such a case the act of investment is a moral act; and he may also feel obliged to choose an investment in which his money is used to finance an undertaking of which he can approve; but nevertheless the act of investment as such is the non-moral and purely utilitarian act of lending money with the expectation of a good return, and to lend money otherwise than this is not properly investment at all. Similarly, a man may feel morally obliged to bring a criminal to justice; but though in that case his act has a moral aspect, it has also a non-moral aspect in so far as the steps he takes in the matter are steps designed simply to bring about a given result irrespective of that result's moral qualities.

All moral action thus has within it a subordinate element of economic action; and there are cases in which this economic element is supreme, to the exclusion of all explicitly moral motives. Implicitly, perhaps, moral motives are never wholly absent. It may be argued, and perhaps rightly, that all actions which on the face of them are purely economic have an implicit moral motive: that, for instance, when I drive a bargain and appear to be thinking only of getting the best bargain, I am in fact restrained, by the force of moral tradition, from adopting underhand methods of dealing which would actually give me a better bargain; or again, I may appear to obey a law solely through fear of punishment, when a careful analysis of my motives would show that I believe it to be right that law in general, or this law, should be obeyed.

On the other hand, just as there are some economic actions in which the moral motive is either absent or at most implicit, so there are actions in which the economic motive is absent or implicit. When a child runs shouting round the garden, it does not aim at any profit or advantage. It feels a desire to act in a certain way, and acts in that way without more ado. The action is, or at least may be, advantageous to its physical or mental development; but it does not think of that. Or it may, by so acting, please or displease someone whose pleasure or displeasure may affect its happiness; but that again does not enter into its calculations. It has no calculations. And just as a person may perhaps be condemned as immoral who acts on economic motives without raising the question whether he is justified in so doing, so a person may be condemned as imprudent who obeys his impulses without reflecting on the economic aspect of such action. Indeed, the reason why one thinks it important to reflect before indulging one's appetites is not merely that such indulgence may lead to immoralities, but also that it may lead to imprudences.

It is thus possible to distinguish three types or forms of action. First, the doing something because it is what we want to do; secondly, the doing it because it is expedient; thirdly, the doing it because it is right. The first is the sphere of impulsive action; the second, of economic; the third, of moral. These three are not mutually exclusive species of a genus. There is no action in which impulse or desire does not play a part, and there is, therefore, a sense in which all action is impulsive. So far, the hedonist, who argues that everybody does precisely what he wants to do, is in the right. But though impulse is, as it were, the foundation of all action, the hedonist is wrong in arguing—as in effect he must argue—that economic and moral acts differ in nothing essential from acts of pure impulse. The mere fact that he has to twist these types of action into conformity with his standard shows that hedonism is a dialectical *tour de force* rather than an unbiased statement of the facts. We applaud his ingenuity in showing that the sweated labourer and the religious martyr are simply enjoying themselves, but even he is not really convinced by it. In economic action impulse is, though present, subordinated to utility, much as in moral action utility is subordinated to duty. Hence the hedonist's effort to drag economic action into line with impulsive action is parallel to, and cannot succeed better than, the utilitarian's effort to drag moral action into line with economic.

The distinction between these three forms of action is not an accidental or arbitrary distinction. It is a philosophical distinction; one which is made not because we choose to make it, but because we cannot think about the subject at all without making it. For the present our concern is with economic action only, and therefore we need not raise the question of whether the three forms of action above distinguished are the only forms whose conceptions are in this way philosophically necessary, or whether a closer examination would reveal others. Our present task is to make clear to ourselves the concept of economic action, and we must leave to empirical economics the question of what particular actions are economic or expedient in particular circumstances, and the attempt to classify, in a manner none the less useful for being arbitrary, these particular economic actions.

In attempting to define the concept of economic action we shall derive help from reflecting on such epithets as useful, expedient, profitable, prudent; such distinctions as those between means and end, profit and loss, debit and credit, receipt and expenditure; such concepts of action as purchase, exchange, and valuation. All these terms have one feature in common: they all assert a relation of a peculiar kind between two things. That is useful which enables someone to enjoy something else which he wishes to enjoy. A prudent act is one in which the agent looks beyond what he is doing and by its means secures something which is other than itself. Purchase is giving one thing in order to receive another, and so forth. This duality is the characteristic feature of economic action. Two elements are present in it; first, the immediate act, and secondly, the mediate act. The immediate act is the means, the expenditure, the price; it is this act which is described as prudent, or useful, or the like. The mediate act is the end, the receipt, the thing purchased; it possesses not utility, but (if one may use the word) desiredness. An economic act is an act which differentiates itself into two parts, one of which is done for the sake of the other. The act is not complete till both these parts are present.

This duality is peculiar to economic action. It is not present in impulsive action, for here the distinction between means and end is simply non-existent. If we take the case of the shouting child, or of an angry man kicking a chair, and try within this act to distinguish means and end, we find no room for such a distinction. It is a misrepresentation of the facts to say that the man kicks the chair as means to work off his anger; if that were a true account of the case, it would not be an impulsive act. A man might perhaps say to himself: 'I feel angry; how can I work off this passion? Perhaps kicking a chair would do it; let us try;' but if he did, it would be a case not of impulsive, but of economic action. In moral action, on the other hand, the distinction is present, but it is merged in a fresh unity; thus it is actually a definition of moral action to say that a moral act is an end in itself, that the good will is the will which wills itelf, and the like. The end which the moral man sets before himself is to be good; and the only means to being good is—being good. Yet the distinction between means and end is here a real distinction, though a distinction without a difference.

The distinction between impulsive and economic action may be stated by saying that in impulsive action we do what we want, whereas in economic action we do what we do not want in order that we may do what we want. That which is immediately desired is the end; the means is only mediately desired, as the necessary presupposition of the end. That which is immediately done is the means; the end is only mediately done, as consequential upon the means. Hence what we immediately do is, in economic action, never what we immediately want to do. This fact is rich in consequences. In the first place, economic action on its immediate side appears as the antithesis to impulsive action. It is doing what one does not want to do, as opposed to doing what one does want to do. It is therefore not pleasant, but painful; it is labour, as opposed to enjoyment. It is not getting what one wants, but sacrificing what one wants; it is thus in its immediate aspect not receipt, but expenditure. And in the second place, the same action on its mediate side appears as a gain or enjoyment, but one which arises only indirectly from our own activity; therefore it appears as something coming to us from outside ourselves, something which we obtain from an external world, something which we earn and do not create. Both these aspects of economic action are illusory, for both disappear when the two sides of the action are taken together. In the light of the end, the means cease to be merely laborious or painful; they become an integral part of that which, taken as a whole, we desire. In the light of the means, the end ceases to be something outside ourselves which is given to us as a result of of our labours; it becomes our own activity regarded as achievement. Thus, a man sweeps a path to earn a loaf; but what he gives is not mere labour, but the labour necessary to earn his loaf, and what he gets is not the loaf, but the act of eating it; or the act of giving it to his children. In so far as he is conscious that the sweeping is the necessary means to the eating, the sweeping itself becomes an object of desire; but it is, even so, only mediately desired, that is, only desired because of the end to which it is the means; and there is always and necessarily a conflict between its mediate desiredness and its immediate distastefulness.

This distinction between the mediate and the immediate sides of an economic act is mythologically represented by the distinction between two persons called the purchaser and vendor, producer and consumer, or the like. 'The purchaser' and 'the vendor' are obviously abstractions, for no one can be purchaser without being at the same time a vendor, and vice versa: every sale is an exchange, and each party sells what he gives and buys what he takes. But they are more than abstractions; they are mythological figments; for there is no such thing as an exchange between persons. What one person gives, the other does not take. I may give a piece of bread for a cup of milk; but what I give is not the bread, but my eating of the bread, and this is not what the other party gets; he gets *his* eating of the bread, which is an utterly different thing. The real exchange is my giving up the eating of bread and getting the drinking of milk; and there is another exchange, that of his drinking against his eating, on the part of the person with whom I am said to exchange commodities. All exchange, in the only sense in which there can be a real exchange, is an exchange between one person and himself; and since exchange, understood as the relation between means and end, is the essence of economic action, all the essentials of economic theory can be worked out with reference to a single person. Indeed, it is only so that they can be intelligibly worked out; for the appeal to a number of persons implies the fallacious doctrine that something exists, called wealth, which can be transferred from one person to another. The reality concealed behind the mythological idea of wealth is the enjoyment of a desired activity; and this cannot be transferred from person to person. The truth behind the idea of transferring wealth is that where a number of persons stand in determinate relations to one another the activity of one is so bound up with the activity of the rest that cases constantly arise in which one person can do what he wants only if someone else does not do what *he* wants.[2] But if I forgo something that I want in order that

---

[2] These remarks about wealth must be taken as applying to all the various ideas

another may have what he wants, I do not, except in a metaphorical sense, give him the thing I want. What I do is to make it possible for him to have something else. And he may induce me to do this by making it possible for me to do something I want to do. In this case each party is using the other as means to his own ends by permitting the other to use him in the same way.

This reciprocal relation of means to other people's ends is the root of all economic organizations, and the basis of all their peculiarities. For instance, there is always a conflict of interest between the parties to a purchase, each trying to buy cheap and (what is the same thing) sell dear. So long as this is presented as a merely empirical fact, it is unintelligible from a scientific point of view and disconcerting from a moral. One is tempted to ask, in the first place, why should this be so? and in the second place, need it be so? Would not our economic life be sounder, more truly productive of real wealth, if it were not so? These questions can only be answered by presenting the conflict of interests as a corollary of the concept of economic action.

Cases certainly occur of purchases which offend our moral consciousness. In these cases the vendor accepts a price which we think he ought not to accept; and we may think it morally desirable that he should obtain a higher price than the minimum for which he will consent to sell. But it is exceedingly important to realize that when this happens we are protesting in the name of morality, not against the size of the price, but against the fact that, *in this case*, a price has been given at all. In other words, we are asserting that *this* case morally demands not an act of exchange, but an act of gift. When someone gives me goods or services, and thereby opens to me certain opportunities for actions which I desire to do, my accepting of these goods or services is an exchange or economic act so far as I induce him to give them to me by giving him other goods or services my own desired use of which I forgo. Now it is for me to determine which of two incompatible alternatives I prefer. If I may either eat my bread or exchange it for milk, I have before me two incompatible

---

which revolve round that of wealth; for instance, the idea of production. Productive labour is not labour that produces a thing called wealth; it is a synonym for economic action, which in its immediacy is labour, and as achieving an end is productive. The epithet 'productive' is therefore otiose; a given action is felt as labour only because it is not an end in itself, but is undertaken for what it produces.

alternatives: eating this particular piece of bread and forgoing the milk, or drinking this particular milk and forgoing the bread. I may choose, in such a case, either on economic grounds or on moral. I may simply ask myself which I prefer; or I may ask myself which action I ought to do. In the latter case, I may decide to drink the milk although I do not want it and do want the bread, because—for instance—the other party to the transaction needs the bread, in my opinion, more than I do, and cannot offer me an attractive price for it. But if so, the act is not an exchange, but a gift. My taking something in return does not prevent its being a gift; for the thing which I receive is not what induces me to act.[3] Hence, in this case there is no purchase; nothing is bought, nothing is sold, and nothing that can be called a price changes hands. But in the other case, where my choice is made on economic motives, there is only one reason that can induce me to exchange, namely, that I prefer drinking this milk (not milk in general, nor even a specified quantity of milk, but *this* milk; and not even this milk in any circumstances, but here and now, with this particular thirst upon me) to eating this bread. If, when I am so disposed, I meet another person who prefers the opposite, and if it is in my power to enable him to eat the bread, and in his power to enable me to drink the milk, then we can do what is called effecting an exchange, which means entering into the condition of reciprocity described above; collaborating with each other in economic actions each of which is essentially an exchange between the agent and himself.

Now because, in an economic act, I am answerable to myself only for driving with myself a bargain which shall satisfy myself, it follows that there is no appeal from this act; there is no way of fixing the right price of anything except by finding out the price which, at the moment of purchase, the purchaser is willing to pay. If I can get a certain end by undergoing certain expense or labour, and if I do not sufficiently want the end to undergo this expense or labour, then, so far as economic motives are concerned, I go without. If the price is a higher price than I am disposed to pay, I do not pay it. No doubt I may change my mind and pay it tomorrow; I may pay it five seconds hence; but that will be because circumstances have altered, and when that has happened the act is a different act.

---

[3] There are cases, such as the exchange of presents at Christmas, in which we give expecting to receive; but this expectation does not convert the exchange of presents into a purchase, or exchange, in the economic sense of the word.

This principle is ordinarily expressed in the theorem that the price of a commodity varies directly as demand and inversely as supply; these two factors being really identical, since the demand for a given commodity is nothing but the supply of other commodities which are to be exchanged for it, and vice versa. Demand, in the economic sense of the word, is not at all synonymous with desire or want; there is no 'demand' for a thing till something is offered in exchange for it, that is, till someone proposes to buy it. The ordinary theorem, therefore, means that prices are fixed by proposals to buy. But this truth is obscured in empirical economics by the idea of quantitative prices, which are supposed to vary as the intensity of demand varies. This idea involves the fallacious assumption that what I pay for is (for instance) bread, whereas what I really pay for is my eating of this piece of bread; and the price of this cannot vary, because it is a thing that only happens once. In strictness, every economic act is unique, and consists of undertaking unique means to a unique end; but empirical economics classifies these acts in arbitrary ways, according to their more-or-less accidental similarities, and within the classes so formed it tries to distinguish by a quantitative scale individual cases whose real differences are qualitative.

Prices are thus fixed afresh by every single act of exchange that takes place. Every such act is a determination of value; and value in the economic sense of the word cannot be determined in any other way. When a man who owns certain goods puts a price on them he is simply forecasting what, in the given circumstances, someone will think it worth his while to pay; and when the circumstances change, he must price his wares afresh. Hence all prices are always ultimately fixed by the 'higgling of the market'; but this is, in practice, concealed by the way in which a skilful vendor will anticipate such higgling and put upon his goods a price which, understanding the mind and circumstances of his public, he knows they will pay. And if people will not pay the price which he has fixed, this shows, not that the public is too stupid or too stingy to pay a reasonable price, but that his calculation of what he could get was at fault. The price which he ought to have asked is the highest price that the public will pay.

Thus, when the fallacious suggestions due to its quantitative language have been guarded against, the formula 'buy in the cheapest market and sell in the dearest' is the definition of buying and selling, that is, of economic action. Because the economic act falls into two parts, the means and the end, the end being immediately desired and the means immediately distasteful or laborious, there is a necessary conflict, within the act itself, between the means and the end. We want the end, but we do not want the means except so far as we are convinced that we cannot get the end without them; and therefore every economic act involves an attempt to evade or minimize the means, that is, to substitute for this act another and less distasteful act which shall procure the same end, or an end the same for practical purposes. This is the conflict of interests which is recognized as subsisting between the parties to any exchange. It is not a conflict between person and person; it is a conflict between the mediate and the immediate sides of economic action as such, and it is quite as prominent in the economic life of a Robinson Crusoe as in that of a retail fishmonger. In this conflict of interests the economic life consists; for the economic life is exchange, and exchange consists in balancing means against end in such a way as to get the best available return for your expenditure.

It is, therefore, impossible for prices to be fixed by any reference to the idea of justice or any other moral conception. A just price, a just wage, a just rate of interest, is a contradiction in terms. The question of what a person *ought* to get in return for his goods and labour is a question absolutely devoid of meaning. The only valid questions are what he *can* get in return for his goods or labour, and whether he ought to sell them at all.

As soon as any moral motive is imported into an economic question the question ceases to be an economic one, and the price, or wage, or interest becomes a gift. But the demand for a just price or a just wage is not a mere confusion of thought. It has two motives, one sound and the other unsound. The right price of anything is what it will fetch at a given moment on a given market; but it does not follow that everything ought always to be on sale in every market. By threatening a man's life or reputation I may induce him to give me money; this money is the price of my forbearance, and the fact that I can get it shows that it is a perfectly fair price. If a blackmailer puts his figure too high, he will be told to publish and be damned. But because in certain circumstances a man will give a thousand pounds for some old letters it does not follow that the circumstances ought to arise. The blackmailer creates a demand for certain commodities, just as the advertiser does, in order to enhance their price; and, from an economic point of view, there is nothing to be said against his doing so. But the particular means which he adopts in order to create a demand are contrary to law, and he is therefore a criminal; or, if you like, contrary to our moral ideals, and therefore we call him a bad man. Now it is always possible, and often desirable, to raise the question [of] whether the circumstances which induce a given person to pay a given price for a given commodity ought to exist. The wages of sweated labour are economically irreproachable, because the labourer is willing to work for them, though (like everyone else) he would get more if he could; but the reason why he is willing to work for them is that he is so poor; and it is very desirable to ask whether, that being so, he ought not to receive gifts instead of wages. It may be held that the bargain which he makes is a bargain which nobody ought to be allowed to make. The notion that there are certain kinds of bargains which must not be made is familiar enough;[4] and it is an equally familiar principle that certain circumstances ought to debar persons from making certain bargains; for instance, tender years or infirm understanding. And it is arguable that excessive poverty should, while entitling a person to receive some kind of maintenance, debar him from offering his labour or possessions in the open market. The demand for a just wage, a wage fixed by legal or moral, rather than by economic, standards, is a rational demand and deserving of respectful attention if it is based on the belief that special circumstances, which ought not to exist, induce certain wage-earners to accept a lower wage than that which they would accept if these circumstances were removed. If such circumstances exist, they ought to be removed, for instance by legislation; and since this legislation would raise some people's wages, the wages as so raised might be loosely described as wages fixed by law. But they would really be fixed not by law but by supply and demand in a market where law insured fair bargaining. In a word, the demand is reasonable so far as it is a demand, not for legislation directly controlling wages—that is an impossiblity, since a wage fixed by

---

[4] A great deal of confusion on this point is visible in current discussions concerning such things as gambling. There cannot conceivably be any economic argument against gambling; if I choose to pay a certain sum of money in return for a certain profit contingent on my luck or skill, I am paying market price for a commodity which I desire, and there is an end of it. But it is arguable that here, as in the case of prostitution, exchanges are made which ought not to be made; and if that is so, the further question arises of whether they can and ought to be prevented by law.

any but economic considerations ceases to be a wage—but for legislation amending the condition of society.

On the other hand, the demand for a just wage is wholly irrational if it is a demand for the control of wages by any standard other than the act of exchange itself. Such standards have often been advocated; it has been argued that the wage earner has a right to a living wage, to the whole produce of his labour, and so forth. Such proposals as these are always in practice 'rationalizations' of a dissatisfaction with existing wages. They do not seriously propound a standard; they only cover a vague demand for more. But the demand for more is an inevitable feature of all economic action; it is impossible that such demands should not be made, and made constantly. Because every exchange consists of buying cheap and selling dear, every exchange is accompanied by a more-or-less explicit wish that one had bought cheaper and sold dearer. Everyone is in some degree dissatisfied with every bargain he makes, and wishes he could have made a more advantageous bargain. This dissatisfaction would be replaced by resignation if he could be quite sure that the price he paid for his end was an absolutely necessary price. But within the economic sphere no one can ever be quite sure of that. If the means and the end were envisaged in such an indissoluble unity the dualism, which is the differentia of economic action, would vanish. Hence to be dissatisfied with one's own bargain is inseparable from the economic life; it is only when we cease to bargain that we cease to be dissatisfied with our bargains.

This truth is intuitively recognized by that common sense which derides as a mere utopian dream the longing for a just wage, a non-competitive economic system, a society in which mutual suspicion shall be replaced by mutual respect, and bargaining by generosity. Conflict of interest, competition, the will to drive a hard bargain, is the essence of economic action, and a society in which economic action has no place is certainly a utopia. It is mere sophistry to argue that, because co-operation is necessary to any economic system, competition is wasteful or economically undesirable; for the presence of these two opposites together is essential to an economic system. The parties to an economic action co-operate in competing, like two chess players.

Yet these utopian dreams, these rebellions against the sordid aims of the economic life, against the worship of gain and the acquiescence in a competitive system, are not wholly to be condemned.

They are both foolish and vicious if they proceed from a desire to enjoy wealth without winning it in the open market. If people who cannot get as high a price as they want for their goods or labour complain that only a ruthless competitive system prevents them from getting more, they are merely throwing a cloak of hypocritical moralizing over their own disappointed greed.The competitive system of which they complain is just the fact that they, and people like them, want all they can get. But the economic life is not everything; and it is right to protest against the assumption that buying cheap and selling dear make up the whole duty of man. Indeed, a renunciation of purely economic aims is the essence, negatively defined, of the moral life.

Exchange, in the sense of exchange with one's self, the balancing of means against ends, is the fundamental conception of economics. The task of economics as a philosophical science is confined to stating this conception and working out its necessary implications; the task of economics as an empirical science consists in showing how this activity of exchange reappears from point to point in a constantly changing historical process; how the social structure of a certain time and place—especially of our own time and place— reveals the features which, according to philosophical or pure economics, all economic activity must present. Whether a conception like that of wages, or capital, or money belongs to empirical or pure economics is a question of utmost importance for pure economics, but one which empirical economics can afford to ignore; for an empirical science reduces every concept to the level of an empirical generalization.

Of the three concepts just enumerated perhaps the only one that belongs to philosophical economics is that of money. Wages are only a special case of prices, and are peculiar only in being the price paid for a peculiar kind of commodity. Capital is a concept which stands or falls with that of wealth, and we have seen that philosophical economics must resolve the idea of wealth into the idea of economic action.[5] But the notion of money introduces a real differentiation

---

[5] Capital is wealth used in the production of other wealth, but not thereby either consumed or parted with. Hence it differs from labour, because labour is consumed, and from a price, because a price is parted with. But *something* is parted with, namely, the using of one's capital for another purpose; and the using of capital is really a mere case of exchanging one action for another, possession of the capital being equivalent to being in a position to choose between these actions.

into the notion of exchange, and we may therefore conclude with a
few remarks on this subject. Money is the means of exchange; that
is, something taken in exchange only to be given in exchange again.
The simple act of exchange, which we have seen to be distinguishable into two correlative acts called the means and the end, is by the
introduction of money made into a compound act, distinguishable
into two pairs of correlatives in which the money, like the middle
term in a syllogism, appears twice over: first we exchange one thing
for money, and then we exchange the money for something else.
Thus money is an interpolated term between the means and the
end, and an economic act into which money enters is a twofold act,
which is complete only when the money has been both accepted and
spent by the agent in question.

Into the practical value of money as facilitating exchange and thus
affecting the organization of society we need not here inquire; that
belongs to the task of empirical economics. But money in its essence,
so far from being a social expedient, plays its part within the
economic life of the individual. When anyone looks forward to an
end and adopts means in order to attain it, the relation between
means and end is never immediate. If the means led instantly to the
end, there would be no appreciable distinction between the two,
and the economic character of the action would disappear. There is
always, in an economic act, a certain mediation, passage, or transi-
tion leading from the means to the end. This passage can be
distinguished into phases, each dependent on the preceding phase,
and therefore it is not a single indivisible act, but a chain of acts
connected by their leading up to the end. The making of a journey
in order to reach a certain place, the daily practising of a difficult
piece of music, the development of the tactical plan of a battle, are
examples. Now these intermediate stages which connect the means
with the end have certain characteristics of their own. They are not
mere means; relatively to the starting-point, they seem almost like
ends. It is a positive satisfaction to see them emerging; even when
they actually include the most difficult and painful part of the
process, there is in them a definite sense of achievement. On the
other hand, relatively to the end, they seem like mere means, for
their only value is that they lead to the end. They have this double
character: that they are aimed at with the deliberate purpose of
passing beyond them. But this is the pure definition of money; and
the empirical fact which we call money is only a special case of such

intermediate stages between means and end, just as the empirical fact which we call purchase is only a special case of exchange, or the adaptation of means to ends.

In the empirical fact which we call money a number of mixed motives are at work, which it is not the concern of philosophical economics to analyse. But so far as what we call money really is money, so far as it fulfils the function of a means of exchange and no other function, its entire character is deducible from the definition given above. It is an end which is also means, neither mere means nor a mere end. If food is an example of an end and labour an example of means, the money with which the labourer is paid and which he gives away for food is neither a kind of food nor a kind of labour; but it has something in common with both. Labour is created by the labourer and is spent by him for the sake of getting something else; food is earned and consumed by the labourer. Money, like labour, is spent, but it is not created; it is earned, like food, but it is not consumed. Money has the peculiarity of being earned to be spent; or rather, whatever is earned to be spent is money. If an article of food, such as salt, is accepted in an act of exchange with the purpose of eating it, it is a commodity; if with the purpose of exchanging it again, it is money. If a bank note is accepted with the purpose of lighting one's pipe with it, it is a commodity; if with the purpose of spending it, it is money. It is always possible for the possessor of money to divert it into use as a commodity at any moment; and where salt, or cows, or cartridges, or lumps of gold are used as money, this is because people want their money to possess an emergency value as a commodity. This may or may not be an advantage to society. If commerce is at a very low level of development, if the possibilities of exchange are restricted and precarious, it is useful to have a currency which can be treated as a commodity in emergencies. Where commerce flourishes, emergencies of this kind are unlikely to arise, and the inconvenience of using cartridges (say) as a currency outweighs the advantage of being able to fire off your money when you run out of ammunition. Hence an advanced, as opposed to a primitive, economic organization is recognizable by its adoption of a currency chosen for its efficiency as money, and not for its emergency value as a commodity. For the same article cannot be at the same time and in the same respect both money and a commodity, and therefore whatever is well adapted to fulfil one of these functions is to just that extent ill adapted to fulfil the other. If you select a certain thing for use as money because of its value as a commodity, you are tempting people to use it not as money but as a commodity. Thus, if children used chocolate money, they would eat it; and if you make a sovereign out of a sovereign's worth of gold, anyone who has a use for gold will melt down his sovereigns instead of spending them, unless you make it illegal for him to do so. And the law which forbids the melting down of coin is an empirical recognition of the truth that money is only money so long as its users pay no attention to its possible value as a commodity.

The same truth is curiously illustrated by Gresham's Law. Why should bad money drive out good? Such a survival of the unfittest seems, on the face of it, paradoxical. It is absurd that if two currencies are in existence, one economically good and the other economically bad, anyone should ever accept the latter instead of insisting on receiving the former; and equally absurd that anybody possessing both should use the bad money rather than the good, if by good is meant good *qua* money, or capable of being exchanged for good value. Since to be money is to be exchanged, the fact that good money is hoarded instead of being exchanged proves that it is not money, but a commodity. The true interpretation of Gresham's Law is that it refers to a state of things in which a commodity, such as cartridges, salt, or gold, passes as currency among people who, in using it, are conscious of its commodity value, and by 'good' money it means money whose face value is equal to its value as a commodity. If now a second currency comes into use, having a commodity value less than its face value, the so-called 'good' currency will cease to be money and will become a commodity. But this proves that to lock up commodity value in currency is wasteful and economically unsound, and that in fair economic competition a currency which is not thus loaded with commodity value will always defeat one which is. The 'best' money, in the sense in which Gresham's Law uses terms, is the worst.

This is directly deducible from the definition of money. Since money, as such, is earned to be spent, not to be consumed, any capacity for being consumed is irrelevant to the efficiency of money as such, and to accept it because it possesses this capacity is to accept it as something other than money. It might be inferred that a perfect currency would consist of perfectly unconsumable objects: gold which could not be made into jewellery, pennies which could not be made into paperweights, and notes which could not be used for lighting a pipe. This is clearly impossible, nor is it necessary. What is necessary is only that people should not, when they accept money, be influenced by the thought that it possesses a commodity value—should not, for example, prefer to accept a kind of bank note that was good for lighting a pipe; and the obvious method of securing this end is—if we may here trench for a moment on empirical economics—to make the commodity value of one's money so small in proportion to its face value that no case is likely to arise in which anyone in using it thinks of its commodity value at all.

The absence of recognized commodity value, however, is only the negative qualification of a perfect currency. The positive side is constancy of exchange value. This again follows from the definition of money. As the middle term of a syllogism must not be ambiguous, that is to say, must not be subject to a change of meaning between the major premise and the minor premise, so, and for exactly the same reason, the mediating link in an exchange must not be subject to an alteration of value between the time when a given person accepts it and the time when he spends it. For he only accepts a given sum of money on the understanding that it will procure him a known quantity of goods; the real exchange is goods for goods, and it is the current price of the goods which he wants that determines the price he will take for the goods he is selling. If he does not know what money will buy, he has no means of deciding how much money to take for the things he is selling; and if he is misinformed as to what money will buy, he is under a delusion in agreeing to accept a certain sum of money. Hence, if people do not believe that the money they accept will remain approximately stable in value until they have spent it, they will not accept it, and it ceases to be money.

These are the only two qualifications of a sound currency. How they are to be realized, or approximately realized, in practice is a question for empirical economics. Whether we want to realize them, or prefer to realize some other aim incompatible with them is a question of politics. For our present purpose it is enough to point out that they cannot be realized except by what is called a 'manipulated' currency. Constancy of value cannot be attained either by using as money a substance whose commodity value is for some reason not subject to fluctuation, or by using a substance whose total quantity is limited by some accident of nature. The values of all commodities are subject to fluctuation, for the value of anything is only the value that a given person in a given situation sets upon it, every act of exchange being a fresh act of valuation; and limiting the total quantity of money in existence does not insure the constancy of its value unless the total quantity of goods available for exchange is also constant. Because in any given purchase the value of the money and the value of the goods for which it is exchanged must be 'equal',[6] it follows that in the sum total of all purchases the same equality must obtain. Hence the only sound currency is one which is 'manipulated' so as to maintain a constant ratio with the amount of goods on the market. This is well known as an empirical fact; everybody knows that if a state floods the market with paper currency, prices rise all round proportionately to the increase in the amount of money in circulation. The same thing would happen if, owing to large new discoveries of gold, a similar increase in gold currency took place. The only serious argument for a gold currency is that the latter event is unlikely, and cannot be arbitrarily brought about, as the increase of paper currency so easily can, by the unscrupulous action of a bankrupt government.

But this argument is based on the fallacious assumption that governments can be kept out of mischief by denying them certain powers, as one denies firearms to a child. In the case of the child the case is simple. But in the case of the government, *quis custodiet*? The people? If so, the people can destroy the government that they distrust and substitute one that they trust. If they cannot make up their minds to trust any, they will be governed by one they do not trust, which is unlikely to respect their economic opinions. A sound

---

[6] It may be worth while to explain what is meant by equality of value. There is, of course, no such thing as 'value', a measurable substance of which equal quantities may somehow exist in a penknife and a two-shilling piece. And if, *per impossibile*, this did happen, no one could have any reason for choosing to have the one rather than the other; for, their values being equal, neither would have any preferability. In point of fact, to the vendor of the penknife, the two-shilling piece is worth more than the knife—he would, of the two, rather have the coin; whereas to the purchaser, the knife is worth more than the two-shilling piece. The value of two things is said to be equal (the knife e.g. is priced at, or said to be worth, two-shillings) when the holder of *each* thinks an exchange advantageous to himself i.e. prefers what the other holds to what he himself holds. This explains the conception of profit, which remains perplexing as long as we allow ourselves to take the phrase 'equality of value' literally. If the values of the goods exchanged are equal, then no one can make a profit except by fraud. But profit simply consists in getting what you want in exchange for what, relatively speaking, you do not want; and hence every true exchange is profitable to both parties.

currency, by creating wealth, would certainly create opportunities for the misuse of wealth. But anyone who advocated an unsound currency on the plea that a sound currency would be a dangerous weapon in the hands of governments might be asked what he thought of allowing governments to handle other dangerous weapons, such as legislative power or the control of armies. A kind of moral cowardice has prompted some people to imagine that in creating machines—and money is a machine—man may become the slave of his machines, may be led, by the mere fact that they exist, into using them disastrously or immorally. Such a fancy is a mere mental vertigo or loss of nerve; it is not amenable to argument; and anyone who is driven by it to this or that form of self-mutilation is, perhaps, better mutilated. That is not our present question. We are asking, not whether we shall establish a sound currency, but what we mean by a sound currency; not whether we are likely to shoulder successfully the responsibilities of wealth, but what wealth is.

3
Goodness, Caprice, and Utility

We are now ready to answer the question: what kinds of things possess goodness? And the answer which we are ready to give is as follows. It falls into four clauses because it involves two distinctions.

A.1. Absolutely considered, everything without exception is good. The world is good as a whole and everything in it partakes of its goodness. There are nevertheless degrees of goodness among these things.

A.2. Relatively considered, that is, relatively to a standard imposed by agents and arising out of their action in so far as that action involves choice, some things (i.e. those whose degree of goodness comes up to the standard in question) are good and others, those which fall short of it, are bad.

B.1. Primarily, action and only action is good. Every act is absolutely good in its degree; and where one act is chosen as satisfying a certain standard of goodness which an alternative fails to satisfy, the first is good relatively to that standard, the second bad.

B.2. Secondarily, anything whatever is good in virtue of its relation to an act that is good. The goodness of whatever is not an action is a secondary or derivative goodness. Thus, a chocolate is said to be a good chocolate because it provides good eating. The goodness of the chocolate is secondary to the goodness of the eating.

We have now to attack a different set of questions. Hitherto we have been dealing only with the question of what goodness is, and the questions immediately arising out of it. Our answer is that the goodness of a thing is the fact of it being chosen i.e. we have defined what at first sight seems to be a quality of things, called goodness, in terms of a specific activity, called choice.

Having thus answered the question concerning the 'nature' of

Extract from 'Goodness, Rightness, Utility: Lectures delivered in H.T. 1940', Collingwood MS, DEP 9, pp. 34–45.

goodness (I borrow the antithesis from Sir David Ross)[1] we pass to the question concerning the 'grounds' of goodness. We know wherein the goodness of a thing *consists*: we now ask what *confers* that goodness upon it. If the goodness of a thing consists in the fact that somebody chooses it, our new question comes to this: Why do people choose what they choose? What are the grounds of choice?

For Professor Moore, I suppose, this would always necessarily be a nonsense question. His doctrine that good is a simple quality like yellow implies, clearly enough, that no significant question can be asked concerning the grounds of anything's goodness; and if a thing's goodness is the same as the fact that somebody chooses it, Professor Moore's doctrine is that there is never any reason why anybody chooses anything.[2]

This is certainly untrue: but it represents an exaggerated and monstrously magnified distortion of a half-truth. I must repeat that the facts which it is our business in these lectures to investigate are facts of consciousness. When I ask, therefore, whether Professor Moore is right in saying that there is never any reason why anybody chooses anything I am asking whether Professor Moore is right is saying that nobody is ever *conscious* of any reason why he chooses anything. The search for reasons is to be conducted and their presence or absence established, under the daylight conditions of self-consciousness, where self-consciousness is accepted as a court from which there is no appeal. We are asking no psychoanalyst to go fossicking round in our unconscious, looking for motives with a lantern as Diogenes looked for an honest man. If any psychologist

[1] The author is here referring to W. D. Ross, the Provost of Oriel, whose book *The Right and the Good* (Oxford, Oxford University Press, 1930) Collingwood greatly admired, not, however, without serious reservations.

[2] Collingwood regarded G. E. Moore as one of the principal Cambridge Realists whose 'Refutation of Idealism', *Mind* NS 12 (1903), was a significant contribution to the advancement of the school, or allied schools, in both Cambridge and Oxford. Collingwood was not impressed by the article which he believed attributed views to Berkeley not entirely dissimilar to those that Berkeley wished to controvert. See R. G. Collingwood, *An Autobiography* (Oxford, Oxford University Press, 1970: first published 1939), 22. George Edward Moore (1873–1958) was a fellow (1898–1904), a lecturer (1911–25), and then Professor of Mental Philosophy and Logic (1925–39), in Cambridge University. He was the editor of *Mind*, 1921–47. The doctrine to which Collingwood refers is articulated in G. E. Moore, *Principia Ethica* (Cambridge, Cambridge University Press, 1903). Moore says, 'My point is that "good" is a simple notion, just as "yellow" is a simple notion; that, just as you cannot, by any manner of means, explain to any one who does not already know it, what yellow is, so you cannot explain what good is.' p. 7, sect. 7.

should intervene in our inquiries with information about the unconscious reasons which people have for doing things, we shall thank
him, politely I hope, but we shall put him as gently as possible
outside the door, saying 'what you have to tell us is very interesting
and we do not question its truth; but it has nothing to do with the
question we are asking—we want to know whether there is anybody
on deck, and if so what they are doing. We are only distracted from
this very simple inquiry by being told that there are various men
below: some in irons, some asleep, some drunk, some gambling,
and so forth.'

My own consciousness being the first consciousness to which I
have access (not of course the only one, because I share with others
the use of certain languages whereby our consciousnesses become to
some extent common property), I ask myself: 'Is it true that I am
never conscious of any reason why I choose one thing and refuse
another?'

I answer 'No, it is not true. It is true that I am sometimes not
conscious of any such reason; but I am sometimes conscious of
reasons; and these reasons are sometimes of one kind, sometimes of
another.'

To develop the first point. There are two kinds of choice (and
remember, please, that when I speak of choice I am speaking of an
agent's consciousness of himself as choosing).

1. There is a kind of choice where the agent is conscious of no
reason why he chooses. He is conscious of having alternative possible
actions before him, and he is conscious of choosing one thing and
refusing another or others; but he is not conscious of any reason
why he does this. His own choice is for him a 'brute fact' (that is, a
reasonless or irrational fact; 'brute' in the sense of 'irrational'). A
choice which to the chooser's consciousness is a brute fact is
ordinarily called a *capricious* choice or a caprice.

Caprice means goatishness; and in one sense goatish behaviour is
perfectly rational. It arises from psychical conditions which can
sometimes be described, perhaps with accuracy, by an intelligent
observer: and the way in which these psychical conditions give rise
to goatish behaviour can sometimes be referred, by a sufficiently
intelligent observer, to psychological laws. Here, then, is a legitimate field for psychological inquiry. Granted that, from my own point
of view, a certain choice of my own is capricious, I am very ready to
grant that a psychologist may succeed in explaining it by reference

to psychological laws; and as a matter of fact I am glad that he should try, and take a lively and I hope intelligent interest in his attempts. This interest is further stimulated by my belief that it is a wise interest; because I believe that in cases of a certain kind it is useful for a man to know something about the workings of his own unconscious and to understand the psychological laws according to which his capricious actions proceed from these unconscious psychical forces in himself. The cases in which I think it useful are cases in which his capricious actions are 'morbid' actions, that is to say actions that interfere with the life he is trying to lead.

For example, I may be trying to live in such a way that I must on certain occasions obey orders from a man whom I regard as my superior. Now, I may find myself on such occasions disobeying such orders; disobeying them as a matter of choice, choosing to disobey them, but disobeying them capriciously, without being conscious of any reason for so doing. I am conscious that what I choose to do is to disobey; in other words, I think disobedience *qua* disobedience is a good thing. The goodness of disobedience, as I see it, is a simple quality of disobedience, a brute fact like yellowness. But this brute fact is a fact which makes me unable to live in the particular way in which I am trying to live; and hence not only I myself in so far as I am trying to live in that way but anybody else who is co-operating with me in the same attempt regards it as a morbid fact. If a psychologist can show that I disobey because I have an unconquerable emotion of hostility towards anyone whom I regard as my superior, and if he can further show that this emotion is due to what, if he is a Freudian[3] he calls an 'Oedipus complex' or if he is

[3] Although Collingwood was extremely critical of psychology's claim to be a science of the mind he nevertheless acknowledged its importance as a science of the *psyche*. See e.g. 'Aesthetic' in *The Mind*, ed. J. S. McDowall (London, Longmans, 1927), 236–7; *Autobiography*, p. 95; *The Principles of Art* (Oxford, Oxford University Press, 1977: first published 1938), 164, 218–9; *An Essay on Metaphysics* (Oxford, Oxford University Press, 1940), 101–42; *The Idea of History* (Oxford, Oxford University Press, 1973: first published 1946), 221–5. Collingwood believed that Freud had made a profound contribution to psychoanalysis, but had over-extended himself in trying to apply his theories to anthropological studies of primitive societies. See [Fairy Tales B] 'II. Three Methods of Approach; Philological, Functional, Psychological', Collingwood MS, DEP 26, pp. 29–44; and *Principles of Art*, pp. 62, 64, 77 [n., 127] n. In this respect Collingwood was critical mainly of Sigmund Freud, *Totem and Taboo: Resemblances Between the Psychic Lives of Savages and Neurotics* (London, Penguin, 1938: first published 1919). Collingwood took psychology seriously enough to undergo a full 50 sessions of psychoanalysis before considering himself qualified to comment on it. See J. D. Mabbott, *Oxford Memories* (Oxford, Thornton's, 1986), 76.

an Adlerian[4] an 'inferiority complex', he has diagnosed the disease which makes me incapable of living in the way in which I am trying to live. If he is able not only to recognize this so-called complex but to remove it, he can cure the disease as well as diagnosing it. If this can be done, as psychologists believe it can, psychologists are extremely useful and beneficent members of society; not because their work makes any contribution to moral philosophy, but because it deals with problems which do not belong to moral philosophy but have an importance and urgency all their own.

There is an obvious corollary to this. If the utility and beneficence of psychiatry depends on the psychologist's ability to discover the psychical forces underlying this or that form of capricious action, and to dissipate these forces, it is of the utmost practical importance that his operations should be confined to cases where the capricious action is of a morbid kind. Like any other member of the medical profession he must cure those who are sick, and not those who are well. And it is for the patient, not the doctor, to decide whether his own health is satisfactory or not. I will allow for the sake of argument that a psychiatrist is able to eradicate any given form of capricious action by detecting and dissipating the psychical forces that produce it, as a surgeon is able to amputate any limb and remove any organ from the human body. It is in fact no harder for a psychiatrist to eradicate a healthy form of capricious action than for a surgeon to amputate a healthy limb: but the two acts are equally criminal and equally incompatible with the first duty of the medical profession. It is true, and this is the truth I have accused Professor Moore of exaggerating, that capricious action is an indispensable kind of action. A loss of the power to act capriciously would be a far more serious loss to a man than that involved in castration (since a man can live without begetting children and all that it implies) or the removal of his stomach (since it is possible to digest food without having a stomach). A psychiatrist who used his powers in order to eradicate healthy types of capricious action would be a criminal of the most pernicious kind. And for everyone except

---

[4] Alfred Adler (1870–1937) was one of the circle, including Freud, Otto Rank, and Carl Jung, known as the Vienna Psychological Society, later named the Vienna Psychoanalytic Society of which Adler became president in 1910. He left the Society in 1911 and developed his own Individual Psychology which emphasized individual responsibility and cognition, rather than reductionist explanations of the unconscious. See J. Corsini (ed.), *Encyclopedia of Psychology*, (New York, Wiley, 1984), i, 15–21.

those who suffer from really serious psychological disease the only sensible advice is James Thurber's: 'Leave your mind alone.'

2. There is a kind of choice which is not capricious: a kind in which the agent is conscious of having reasons for choosing one of the alternatives before him and refusing the other. This is a choice based on reasons and is called rational choice.

I have already called your attention to Professor Moore's implicit denial that there is such a thing as rational choice, and his implicit assertion that all choice is capricious. I do not for a moment believe that he would say this outright if the question were put to him, but he does say it implicitly; for when he compares the way in which we recognize something as good with the way in which we recognize something as yellow he is surely implying that the two recognitions are alike in this, that precisely as we ask for no reason why we should think a thing yellow, and could not give a reason if one were demanded, but could only say 'I see it yellow and there's an end of it', so we ask for no reason why we should think a thing good, and could not give a reason if one were demanded, but could only say 'I think it good, and there's an end of it.' Now this is true of capricious action, but not true of rational action. It is, indeed, the definition of what distinguishes capricious action from rational action. Thus Professor Moore has not so much denied the existence of rational action as assumed its non-existence, and has built his moral philosophy on that assumption; and herein, I will venture to add, he is characteristic of the so-called Realism of which in this country he is the acknowledged leader; for this realistic movement is essentially an irrationalist movement: a crusade against reason in all its forms, sometimes using the weapons of argument, sometimes (more wisely) simply ignoring the thing it wishes to discredit. Professor Moore has publicly described his own life's work as 'A Defence of Common Sense', and 'Common Sense' by long-established usage is the accepted name for low-grade thinking, theoretical or practical; thinking below the level of reason, below the level of science, below the level of criticism or justification: the kind of thinking which is content to think 'this is so', and when asked for a reason replies, 'this is so because it is so'. In the case of practical thinking, 'Common Sense' is caprice; and therefore it is not surprising that the arch-Realist, the self-styled defender of common sense, should assert that all choice is capricious choice.

Rational choice is the category under which an agent is conscious of himself when, in addition to being conscious of himself as confronted by alternative possibilities for action and also of choosing between these alternatives, he asks himself 'Why do I choose this one?' and gives an answer. The answer to the question 'Why do I choose this alternative?' is, of course, also an answer to the question 'Why do I call this alternative good?' and, as involving an answer to the question 'What is there about this alternative in virtue of which I choose it', involves an answer to the question 'Why is this alternative good?' These four questions are so connected together that we may conveniently regard them as one single question. And the form of words in which we shall find it most instructive to state this single question will be, I think, the following: 'Why do I choose this?'

The infinitely various answers which are given to this question may be grouped into various types, and what I now have to do is to enumerate the chief types. I shall begin by distinguishing between two types of answer, mutually exclusive and at the same time exhaustive, so that any possible answer falls under one or the other type. These I will call *Unreal Answers* and *Real Answers*. By a real answer I mean one that really answers the question: by an unreal answer one that does not really answer the question. You may think this a silly way of dividing the possible answers to a question: but I defend myself against the charge of silliness by pointing out that people very often give unreal answers to questions because, through not thinking carefully enough what they mean, they mistake unreal answers for real answers. A real answer to a question in the form: 'Why is this thing A?' will be the form: 'Because it is B, and B implies A.' An unreal answer to a question in the same form will be in the form: 'Because it is A.' The unreal answer is not a real answer because it is tautologous.

Now, people do not give tautologous answers outright, as a rule, except in order to convey the idea either that they have no answer to give, or that they do not here and now propose to give one. An unfortunate child often gets tautologous answers. 'What are you doing?' 'Writing.' 'Why?' 'Because I am.' This might be said as implying, 'I don't know why I am writing,' or as implying 'I am too busy to explain.' The person who gives a tautologous answer outright knows that he is not giving an answer at all. But people often give a disguised tautologous answer, failing to see through the disguise because they do not think carefully enough what they mean, and consequently believing that they have given a real answer.

Disguised tautologous answers to the question 'Why do I choose this?' may consist of complete or incomplete disguised tautologies. [An] Example of a complete disguised tautological answer would be: 'I choose it because it is good,' or 'Because I think it good,' or 'Because it seems good to me,' or 'Because it seems good to persons with whom I agree.' The tautology consists in the fact that 'good' means the same as 'chosen', so that the answer to the question 'Why is this thing *A*?' is given in the form 'Because it is *A*.' The completeness of the tautology consists in the fact that 'good' means neither more nor less than 'chosen', so that the identity of *A* with *A* is complete. The disguise consists in the fact that the answer is given in good faith by someone who does not know that 'good' means 'chosen'. Through failure to reflect on his own meaning when he uses the word 'good', he thinks he is using [it] to mean something else, though he does not know what.

[An] Example of an incomplete disguised tautological answer would be 'Because it is pleasant,' or 'Because I like it,' or 'Because I expect it to turn out pleasant,' or 'Because I think I shall like it.' The tautology consists in the fact that pleasure is a constituent of activity, a presupposition of choice. The incompleteness of the tautology consists in the fact that it is only one among a number of such constituents or presuppositions. The disguise consists in the fact that the speaker, having failed to reflect on his own meaning when he uses the terms, is unaware of this.

Tautologies are recognizable by the fact that they are not informative. They are not propositions because they propound nothing. No hearer to whom they are uttered can acquire any knowledge from them. No speaker in uttering them is stating anything true, or even anything false. Yet they are not nonsensical. They do not state incompatibles, as nonsensical utterances do. They state nothing at all. Logically they are valueless. They are not necessarily without another kind of value, namely rhetorical value. 'I love you because I love you,' the 'woman's reason,'* does not serve any logical purpose; it does not convey to a man any idea of why the women

---

* Shakespeare, *Two Gentlemen of Verona* I. ii. 24
[JULIA Why not on Proteus, as of all the rest? LUCETTA Then thus,—of many good I think him best. JULIA Your reason? LUCETTA I have no other but a woman's reason; I think him so, because I think him so.]

loves him or why she thinks she loves him; but it may serve to convince him that she really does love him, by showing him that she knows what love is. No one would speak like that, except a person who knew that love is not a thing for which reasons can be given; and I suppose that every lover has asked himself why he loves the person he does love, and has found that there is no answer.

Of real answers to the question 'Why do I choose this?' It will be useful to distinguish three types; (a) 'Because it is useful.' (b) 'Because it is right.' (c) 'Because it is my duty.' I shall try to show, in the first place, what each answer means; secondly, that they mean different things, and thirdly, that they are not incompatible.[5]

(a) The answer 'Because it is useful' may be expressed in many alternative ways. For 'useful' one can say 'expedient', 'profitable', 'prudent', and so forth. And, however it is expressed, it implies a relation between that which is useful and that for which it is useful. The useful is a kind of good whose differentia is this relation. To be useful is one way of being good, namely being good for something. As so related, what is useful or good for something is called 'means' to that something: what it is good for is called its 'end'.

Both means and end are good: that is, chosen. It is only because the end is chosen that the means are chosen as means to do it. Unless the end were good the means would not be good; for the goodness of means is derivative from the goodness of the end for which they are good, and dependent upon it. Nothing can be good for something unless that for which it is good were good simply. Cases occur, no doubt, in which A is chosen as means to B, and B as means to C; and there may be still longer series in which every term is chosen as means to the next; but unless in the agent's consciousness (which is all we are concerned with) the series came to an end somewhere, in an nth term which he envisages as simply good, or good in itself, the series would never begin. The terms up to the $(n\text{-}1)$th would have no end from which to derive their own utility or goodness as means, and would therefore possess no utility. In other words, the agent whose reason for choosing something because it is useful must, when he thus chooses it, already have in mind some end which he serves by choosing it. In the temporal order he proceeds from means to end; in the logical order he proceeds from end to means.

[5] Only (a) is discussed in this extract, but (b) and (c) and their relations to each other and to (a) are discussed in subsequent articles and extracts in this book.

Wherever a case of utility occurs, the agent in whose life it occurs chooses or accepts both the end and the means: he chooses or accepts the end for its own sake, he chooses or accepts the means for the sake of the end. The case which, when we consider that which is chosen, appears as a case of utility or as means and end, appears when we consider the agent as a case of double choice. The two choices are not separable or independent choices, they are logically related as means and end. The end depends on the means in the sense that it is conditional for its realization on the means; the agent cannot have the end without having the means. The means depend on the end in the sense that the means are conditional for being good, that is, for being chosen, on the end's being chosen. The agent only chooses the means because he chooses the end; his choice of the means is a choice logically dependent on the choice of the end.

These two choices, therefore, together make up one single choice. You can call them, if you like, two choices connected by a relation of reciprocal dependence; or if you like you can call them one choice involving two distinguishable, but not separable, parts or aspects. What is chosen, as we have already seen, is always primarily an action: a choice is always in the first place the choice to do something, and only in a secondary and derivative sense the choice of something connected with that action. It follows that the two choices which make up one simple choice are choices of two actions which together make up one simple action. And anyone who likes to verify this inference by appeal to experience will find that it is true. He will find, on examining instances, that when a man has begun taking steps which constitute the means to a given end, the activity which is the end does not lie before him as something wholly in the future, something merely expected or hoped for; it stands in his consciousness as an activity upon which he has already embarked, to which he is already committed. Not deeply enough committed, perhaps, to be guilty of it in the eyes of the legal system under which he lives. If he has bought poison in order to poison his wife, the law may hold that he is innocent of murder until he makes an attempt actually to administer it; but morally, as he very well knows, he is already guilty as soon as he buys the poison; guilty not only of murderous thoughts and murderous desires, but guilty of a murderous action, begun though not yet finished. The degree of guilt certainly differs according to whether the murderous action is finished or only begun. Why some legal systems hold that the degree of guilt involved in this incipient stage of a murderous action is not liable to punishment is a question we need not raise. It is enough to point out, first, that there is no *a priori* reason why the law should take this view, for the law might hold that the possession of means to murder was evidence of murderous intent; and secondly, that our own legal system does recognize this same degree of guilt in respect of other crimes: for example, it recognizes 'loitering with intent to commit a felony' and severely restricts the purchase of 'dangerous drugs'.

The reason why actions are thus divided into means and ends would seem to be that every action is a complex of actions. There is no such thing as an absolutely simple and indivisible action. In some cases, it is true, an agent is hardly conscious that his action is complex: he thinks of it as a simple indivisible whole; and in these cases the distinction between means and end does not occur to his mind. The more he is conscious of the complexity of an action, the more he thinks of it as falling into two distinct phases, logically connected by a relation of mutual dependence, but distinct in the sense that one phase is the completing or crowning phase, the other the preparatory or preliminary phase. These two phases are again subdivisible, but we need not enter into that. The crowning phase is the end; the preliminary phase is the means. The distinction of means and end is thus a distinction arising in the agent's consciousness through reflection on his action's complex structure. Just as the idea of goodness arises through reflection on the way in which his actions begin, starting from a phase in which, because he has not yet begun to do a certain action, the action is present to his mind as a mere possibility, and therefore as one possibility among others between which he has to choose; so the idea of utility arises through reflection on the way in which the action he chooses to do develops through a preliminary phase into a crowning phase. The action proper is the crowning phase or end, but this end, as it presents itself to his consciousness, is an action of such a kind that he can only reach it by passing through the preliminary phase, or means. Hence, when he begins to do the action, he is already conscious of himself as beginning to do the action which is the end: but he is doing it not immediately but mediately, doing it by doing that which presents itself to his consciousness as the means to it.

It would, therefore, be a superficial analysis of the means–end

relation to describe it in terms of time-sequences. The means are
preliminary to the end, strictly speaking, not in the sense that the
means constitute an action which has to be done at an earlier time
in order that, later on, the action which is called the end may be
done; but in the sense that the means-action and the end-action are
simultaneously performed. The means-action is what the agent is
immediately doing (that is, conscious of himself as immediately
doing); the end-action is what at the same time he is mediately
doing, doing by 'means' of the immediate action which we have
called the 'means-action'.

Where an example of the means–end relation is described in terms
of time-sequence, the analysis is always incomplete. This is the case
with the example, given above, in which a man buys poison at one
time as a means to murdering his wife at a later time. Buying the
poison does not constitute the entire means to the murder. Let us
go on. When he gets home he finds tea ready and his wife not yet in
the room. He pours out two cups of tea, puts poison in one cup,
and takes the other himself. Putting the poison in the cup is a
complex act consisting of several manual movements: grasping the
poison, placing his hand over the cup, and letting go. These are all
immediate actions, that is, he is not conscious of any means by
which he does them. The same is true of taking the other cup and
drinking it himself. If we try to discover any one immediate act or
any one set of immediate acts which is or constitutes the act of
murdering his wife we shall not succeed. The truth is that he
murders his wife by means of poisoning one cup of tea and taking
the other. The murder is not an immediate act, led up to by other
immediate acts which are its indispensable forerunners in a time-
series. The murder is a mediate act which is done by means of a
certain set of immediate acts. The mediate act is done simultaneously with the mediate ones which are means to it.

The simultaneity of means and end in time will perhaps be clearer
if we take a simpler case. I wish to illuminate the room. In order to
do so I depress a switch. Depressing the switch is my means;
illuminating the room is my end. There is no time-sequence in
which the means-action precedes the end-action. I do the end-action
by doing the means-action—they are actions which I do simultaneously—I do them both at the same time, but do them in
different ways. The means-action is what I do immediately; the end-
action is what I thereby achieve.

It may still be said, in a perfectly normal sense of the words, that the means-action is done 'first' and that the end-action is done 'after' it. But the terminology of sequence, priority or posteriority, has not here its temporal meaning. We all know that priority and posteriority have a logical meaning as well as a temporal meaning; but here they have not their logical meaning either, because logically the end is prior to the means. The means are chosen only because the end is chosen. The goodness of the means is a secondary and derivative goodness, logically dependent on the logically prior goodness of the means. The sequence of prior and posterior has here a third sense, in which the first term of a sequence is the immediate and the next is said to be after it, or more remote than it, meaning that it is reached mediately through the first or by means of it. There are other examples of the same sense. If I am asked what clothes I am wearing, I reply, 'well, first I have on a vest; then after that a shirt' and so on. The question is not in what order I put my clothes on when I dressed in the morning, but in what order I have them on now. My vest is 'first' in this order because I have it on immediately, that is, it is next to my skin. My shirt is next in order after the vest, in the sense that its contact with me is mediated through my vest: the vest is next to my skin, the shirt is next to my vest. And this is not a disguised or confused way of describing the temporal sequence of my operations in dressing, because it applies to cases (e.g. in anatomy) where a number of integuments have never been put on, but have grown, and grown simultaneously. Thus Professor [Wilfred Edward] Le Gros Clark in his recent book on *The Tissues of The Body* (Oxford 1939) describes the various strata of the epidermis, beginning with this sentence [*sic*]: 'The first stratum in the layer of Malpighi.' In the same sense it is said in the theory of knowledge that sensation is prior to thought.

The utilitarian answer to the question 'Why do I choose *A*?' can be stated, therefore, as follows: 'Because I choose *B*; where *A* and *B* are both actions, and therefore both chosen as good, but good in different ways. *B* is chosen as good in itself, *A* is chosen as having the secondary, derivative, or dependent kind of goodness which we call utility, expediency, or the like: it it chosen as good for *B*. They are also done in different ways. *A* is done immediately, *B* is done mediately. In other words I do *B* by doing *A*; that is the only way in which I do *B* at all; but there is no other action by doing which I do *A*. An anatomist might say that I depress a switch by contracting certain muscles while relaxing others; but I am not conscious of this; so far as my own consciousness is concerned, the downward pressure of my hand on the switch is not done by any means whatever, it is done immediately.' This is the means–end analysis or utilitarian analyisis of goodness.

4
Political Action

Political theory is generally conceived as the theory of the state. The state is regarded as a thing possessed of certain attributes, collectively called its powers or the like. The task of political theory, then, is to scrutinize this thing and give an account of its attributes.

A theory so conceived moves within the four walls of the category of substance and attribute. The state is a substance, an existing entity, and sovereignty, or whatever word one uses to indicate the sum total of its powers, is its essential attribute. A deductive political theory may work out, *a priori*, the implications of this essence; an inductive theory may collect information about the various attributes of sovereignty found to subsist in various states; but all such theories, deductive or inductive, are agreed in accepting the limitations of the category of substance and attribute.

In this paper I propose to approach political theory from a different angle. Instead of putting the central issue in the form of the question 'What are the attributes of the state?' I propose to put it in the form of the question 'What is political action?' That is to say, I propose to take my stand, not on the category of substance and attribute, but on the category of action.

My reason for making this proposal is not any lack of respect for the category of substance and attribute (I would as soon speak disrespectfully of the equator); nor any intention to disparage the theories of the state that are based upon it. Theories of the state, both empirical and *a priori*, are indispensable to any political theory, and everything that I shall say in this paper will presuppose a reasonable acquaintance with them. But all theories of the state, as it seems to me, find themselves confronted, sooner or later, with certain questions that they cannot answer; and these questions are all such as concern the limits of the state, and its relations with other bodies, be they states, or churches, or trade unions, or municipalities. These are the very questions which, alone of all the

Proceedings of the Aristotelian Society, 29 (1928–9), 153–76. Reprinted by courtesy of the Editor of the Aristotelian Society.

questions in political theory, are today matters of urgent practical importance; and, with the example of the great political theorists before us, we need not apologize to each other for thinking primarily, in our philosophies, of the things that matter most in our practice.

Now, it seems to me that these are the questions which, insoluble in terms of substance and attribute, are soluble in terms of action. If you think of politics as a thing, the state, or a number of things, the various different states, having this and that attribute, then the mere 'thinghood' of the state, as so conceived, implies a rigidity of conception, an intransigence of behaviour, which gives rise to a simple destructive dilemma. If the state gives way to the trade unions, or signs the covenant of the League of Nations, *either* it is surrendering its sovereignty and therefore ceasing to *be* a state, *or* it is merely showing that it never possessed such a thing as sovereignty, and therefore ceasing to *pretend* to be a state. In either case, the state is a discredited superstition. This dilemma cannot be undermined by picking a quarrel with the word sovereignty. The fact for which that word stands is the right of the state to do its own work, the work of legislating and enforcing its laws; and it is precisely this work that is taken away from it, not in whole, of course, but in part, and in a part that may well be essential to the whole, by giving in to the trade unions or signing the covenant.

I am not here concerned with the attempts of established political theories to cope with this dilemma. If I were satisfied with any of them I should not be reading this paper; if I laid before you my criticisms of them I should not have time to offer you my own positive suggestions. I will only remind you that modern political theories fall into two main groups—a centrifugal or pluralistic theory which destroys the unity of the state by breaking it up into a number of special associations, and recognizes nothing that it can call the state except one among these associations; and a centripetal or monistic theory which destroys the diversity of the special associations by identifying them all with the state. Each theory has its strength and its weakness. The centrifugal theory tries to do justice to the special associations; and there is this to be said for it, that in an orderly and well-governed country, where the need of a strong executive is not felt, it can do little harm. The centripetal theory stands for the important truth that special associations can only enjoy a successful and active life under the shelter of the state;

and it is the natural theory for a people that has felt the burden of anarchy. But that is the best I can say for these theories, though, naturally, I could say a great deal more in praise of the work done by people who profess them.

Both types of theory fail, I suggest, because they begin by thinking of politics in terms of the state, and never really reconsider this initial assumption. The centrifugal theory finds an active and unmistakable political life existing elsewhere than in the state, and jumps to the conclusion that the state is unnecessary, except as a mere association like the rest. The centripetal theory finds the state compelled to take upon itself other than purely political functions, and jumps to the conclusion that its attributes embrace not only the political functions of society but all functions whatever. If, on the other hand, you start from the conception of political action, and think of the state not as a thing but as the collective name for a certain complex of political actions, the dilemma and all its attendant theoretical difficulties simply disappear. I do not say that the *practical* difficulties disappear. The question whether or not to give in to the trade unions, or sign the Covenant, is, of course, not answered by any political theory except a quack theory. The business of sound theory, in relation to practice, is not to solve practical problems, but to clear them of misunderstandings which make their solution impossible.

Political action is a specific kind of action, resembling in this way economic action. We are accustomed by now, thanks to the work of economic theorists, to the idea that there is a specific kind of action, economic action, aimed at a specific kind of good, the economic good, whose technical name is wealth. The activity of a business man or a business firm *as such* is pure economic activity: it is devoted entirely to the attainment of wealth: and this remains true in spite of the fact that every man is more than a mere business man, and every organization more than a mere business firm. The economic good is so far distinct from the moral good, that in deciding 'this will pay me' the business man is by no means saying 'this is my duty'; and he may, as a moral agent, decide to forgo profits which, had he been a mere business man, he would have tried to procure. But the economic good, though distinct from the moral good, is not opposed to it. A pedantically moralistic view of life may fancy that it is; and for centuries this seems to have been the view of orthodox Christian ethics:

My theme is always oon, and ever was:
Radix malorum est cupiditas.

I need not, I hope, enlarge upon the gain that has come to all departments of human activity through the general realization of the fact that economic activity is not, as such, sinful, but is the pursuit of a real good, a specific kind of value. But it is relevant to my subject to point out that this realization—the clear grasp of the distinction between moral and economic good—only dates from the last two hundred years. And in the case of politics, the realization has not yet come about. Just as orthodox morality once used to shake its head over the organized cupidity, the rampant self-seeking, of the business man, so nowadays it shakes its head over the politician; and as Chaucer's pardoner knew that he could always please his audience by denouncing cupidity, so the modern journalist knows that he can always tickle his reader's sense of superiority by declaiming against politicians. In both cases there is a good deal of hypocrisy in the denunciation; but there is also genuine confusion of thought. It is the merit of economic and moral theorists in the last two hundred years to have cleared up the confusion about economic action; but the confusion about political action still remains. It is traceable not only in the more vulgar forms of journalism, but even in the works of some political theorists, in so far as they try to reduce political problems to economic or moral problems, and thereby show that they do not recognize the existence of specific political values.

Political action, as such, is not moral action. A society makes a law not because it is its duty, but for some other reason—a political reason. Again, political action, as such, is not economic action. A society makes a law not because it will thereby become more wealthy, but because it will thereby achieve a good of another kind—a political kind.

What, then, is the political good? If we can define that, we can say that political action is action aimed at achieving it. But this method is open to objection. The moral good, I believe (you will excuse me from arguing it now), is not different from the moral action that realizes it. Nothing has moral value, I submit, except the will of a moral agent. If that is so, the moral good and moral activity are not related as end to means: they are identical. To do good and to be good are the same. Again, the economic good is wealth, and the economic activity is exchange; for, I find, production and consumption come into the subject-matter of economic theory only so far as they are subsidiary to exchange or actually cases of exchange; and value, in the economist's sense of the word, means exchange-value. If, therefore, to be wealthy is to have things of value, and value means value in exchange, we get the result that wealth must be defined in terms of exchange, not vice versa; and ultimately I do not see what an economic theorist can mean by being wealthy, except exchanging a great many goods. To deny this would involve the denier in elementary economic blunders like asserting that fresh air is wealth to a man on a desert island.

The political good, then, is likely on these analogies to be definable only in terms of political actions, and in a vague and tentative way one may describe political goodness as the goodness of a life which is lived under good laws well administered. But what are *good* laws? Not laws which it was the legislator's duty to pass, for the legislator, as a human being, may be called by duty to a wife's bedside and prevented from passing an act which, none the less, would have been a good one. Not laws which improve the moral state of the subjects, for admittedly bad laws may be an occasion and stimulus to morally admirable actions. A good law is a law which is good in the political sense; if it is also good in other senses, so much the better, but that is not essential. Essentially, a good law is one which achieves the political good, not any other good. But if the political good is a life lived under good laws, and good laws are those which achieve the political good, are we not involved in a vicious circle?

In a circle, no doubt, but not a vicious one; because we are not syllogizing or defining; we are only offering a vague and tentative description. The solution of our circle lies in the observation that the goodness of a law consists in its being really a law—that is, a principle really worked out in thought so as to apply to a particular region of practice, really laid down as binding within that region, and really obeyed or observed within the limits of its application. Until all these conditions are present the law is not a real law at all; it falls short of genuinely being a law; but when they are all present, the law is a good law, so far as it goes. Political action, then, is the making and obeying of laws, with all that this implies: political goodness is the peculiar kind of goodness which this activity as such possesses. This, once more, is not a formal definition; only a revised form of our tentative description.

To get closer to the idea, let us look at a few instances in which the specific form of value, which I call political goodness, is clearly distinguishable from other forms.

Suppose a man wishes to improve the housing conditions of the poor in a certain district: to clear away slums and build better houses. What good does he think this will do. He may, when he asks himself this question, reply that it will make the poor happier. But if happiness is a matter of feelings, of cheerfulness, and jollity, and avoidance of worry and pain, the answer is that very likely it will not. Everybody knows, who has any experience of people in slums, that on the whole they are jollier and less worried than people in garden suburbs. Well then, he may say, it will make them better: it will diminish drunkenness and promiscuity, and lead them into a stricter observance of their duties. But if one judges morality by any but the most bat-eyed conventional standards, one cannot help seeing that the moral state of the comfortable middle classes contains quite as many shortcomings as that of the very poor. If wealth could really put an end to wickedness, it would be an even more enviable thing than it is. Well then, will it make them healthier? That might seem incontrovertible until one collects statistics; but some years ago, when one of the greatest Northern towns wanted to sweep away its worst slum area, an incautious medical officer compiled a table of vital statistics which showed that this area happened to be one of the healthiest in the town.

Ought we, then, to refrain from improving the housing conditions of the poor? I reply, certainly not, but we ought, if possible, to improve them for the right reason: namely, that slums are an evil in themselves, as representing an element of disorder and corporate slatternliness in the body politic, quite apart from any further evils that may come of them. A well-ordered state would not tolerate them, not because they are a danger to happiness or health or morals, if indeed they are, but because they are an offence against order.

Take another case. In England, a crowd of people, all wanting to buy tickets for a show or a train, forms itself into a queue, as it were. by instinct, and bitterly resents the action of anyone who forces himself into the front of the queue, instead of taking his place at the end. What is the good of this singular practice? I call it singular, because I have not seen it in force in any other country, and I well remember my surprise when first I found that in France you had to fight all comers for your turn at the *guichet*. I submit that you cannot give a utilitarian explanation of the queue habit. It does not save time; it is definitely opposed to the interests of all the strongest and most active individuals; and it demands an extraordinary degree of discipline when you are at the end of the queue and the train is about to start. The value which is subserved by it is a political value: orderliness, regularity, submission to a rule which applies equally to all persons.

A third case: I make a promise, and afterwards feel that it binds me, even though neither I nor anyone else can be proved to gain by my keeping it. Why do I feel myself bound? Because it is a duty to keep promises? A circular explanation at best; at worst, a mere untruth. It may be a duty to break a promise. But the keeping of promises, as such, has what I call a political value: its goodness is the goodness of orderly conduct, observance of rule: for a promise has something of the nature of a self-imposed law.

A fourth and last case: suppose you came upon two strangers, one of whom seemed about to shoot the other, who was cowering in a corner. I do not say you would necessarily interfere. You might be afraid, or for various reasons you might allow events to take their course. But one possible action for you would be to interfere and disarm the assailant; and one possible motive for this would be the feeling that one can't have this sort of thing going on . . . shooting people like that . . . This motive is political. You are recognizing, or making, a law against 'this *sort* of thing', and constituting yourself the executive to uphold it. Acting quickly, your thought has not time to mature; you could not, probably, give a satisfactory answer to the question [of] how you would define the sort of thing you object to. But as soon as the thought appears at all, however tentatively, it appears as a generalization. You are not merely reacting in a particular way to a particular situation: you are reacting in a way of a special kind, appropriate to a special kind of situation. There is an element of universality in the particular and essential to it. As in the case of the slums and the ticket-queue, and so forth, it is the principle of the thing that you are objecting to, or insisting upon, as the case may be. The particular case is seen in the light of a general principle. If this general principle were not present, the political value of the act would disappear.

One thing, however, need *not* be present: namely, any answer to the question '*Why* can't you have this sort of thing?' No answer to this question need actually be before your mind as you act: and if the question is put, you may be seriously taken aback. You may try to give various answers and find that they will not do. You may begin by saying 'because it is very wicked to shoot people', but it can easily be shown that this need not be more wicked than other acts which you do not interfere with. You may say 'because he has done nothing to deserve it'; but how do you know? The victim may be a detestable person who has goaded his assailant to the verge of madness by insults and injuries, but, in spite of that, you would still say that you can't have this sort of thing going on. You may say 'the death of this man would be a waste of human life', but you are not entitled to say that until you know what he is good for. In the long run, no answer to the question can be complete. 'I won't have it,' is a declaration of purpose which, however much it may be influenced by reasons and calculations, goes beyond them and is not, in the last resort, based on them. It stands upon its own legs, justifies itself. It is immediate.

In economic action this immediacy is not present: or rather, it is present, but it is broken up into two separate parts, the means and the end. The two together form a single act, which, taken as a whole, is certainly immediate: I pay a penny—why?—to get a bun: but why do I want to buy the bun? I may not want to eat it; and sometimes people buy things when they have no notion what they want them for. The total act of purchase is thus immediate, and not, in the last resort, based on calculations of further ends to achieve. But the purchase necessarily falls into two parts, a giving up and a getting, and this duality is the essence of economic action. People sometimes betray a failure to understand political action by assimilating it to economic action in this respect. If I take a bun and pay my penny, economically speaking, my action is perfect. But if the penny is a fine payable for stealing the bun, then, when I steal the bun and pay my fine, my action remains politically bad; the fine does not pay for the breach of the law; I remain a law-breaker. To suppose that the fine pays for the broken law, as the penny pays for the eaten bun, is to forget that very real thing which is pompously called the majesty of the law—the fact that a command is issued to be obeyed. This fact may be differently expressed by saying that there might be laws with no sanctions attached to them at all; and in effect there are plenty of such laws, for we all habitually obey laws of whose sanction we are ignorant, and commands issued by people who cannot enforce them.

I am suggesting that political action is essentially regulation, control, the imposition of order and regularity upon things. This may be illustrated from the facts ordinarily called political, where someone called a sovereign issues commands to regulate the lives of people called his subjects, or where a community acting through a legislative organ makes for itself laws which, acting through an executive organ, it proceeds to enforce upon itself. But I have tried to point out that the same thing may be illustrated from cases in which a man interferes with the actions of another man in a quite spontaneous and unreflective way, compelling him to obey a law which, on the spur of the moment, he takes it upon himself to promulgate and enforce. It may equally be illustrated from the way in which, by making promises and resolutions, a man imposes laws upon himself. I might also have illustrated it from the way in which we all, quite unreflectively, give commands to children.

It is one of the common errors about political action to suppose that 'the ruler' rules for his own personal benefit.[1] The reason why this mistake is made is that, if ruling means giving commands, we do often give commands for our own personal benefit: for instance, when we tell a cabman where to go, or make a rule that the children are not to blow tin trumpets.

But it is clear that a command of this kind is not intended to bring about an orderliness which we value for its own sake. Therefore, it is not strictly a political command. The political good is order as such; and if we made a rule that the children are always to fold up their clothes, with the purpose of inculcating regular and tidy habits, that would be a genuinely political action. Where a command is given in order to subserve the interest of the person commanding (or, indeed, of any other person), it does not cease to be a command, in other words, it does not part with all its political character, but it subordinates the political good to a good of another kind.

---

[1] The *locus classicus* is, of course, Thrasymachus's position in the first book of Plato's *Republic*. But the fallacy reappears e.g. in the commonplace of socialism, that government by a certain class is *eo ipso* government in the interests of that class. What is really wrong about government by a class is not that it governs in a partial interest, but that it governs with a partial conception of the interest of the whole.

There is a further and deeper reason why it cannot be plausibly maintained that the ruler rules for his own benefit. This is the fact that, as we have seen, a man may regulate *his own* conduct, in which case all the essentials of political action are present, but the distinction between a ruler and a subject is absent. This case, in which control takes the form of self-control, is no mere freak, as it must appear to a political theorist wedded to an Austinian view of sovereignty. It is the fundamental and crucial case of all political action.

Political action must remain unintelligible as long as it is conceived exclusively in terms of ruler and subject. That conception leads inevitably to the question, 'What right have I do interfere with the actions of other people?' and by its very presuppositions that question is unanswerable. But, as Plato saw—the whole argument of the *Republic*, and, indeed, of Aristotle's *Politics*, is based on it— the fact that a man can control others is rooted in the fact that he can control himself. The relation between ruler and subject thus becomes not at all, essentially, a relation between one person who commands and another who obeys, but a relation between something whose nature it is to rule or regulate or introduce order, and something whose nature it is to have order introducible into it— what Plato called Reason and Desire respectively. So far as issuing orders goes, the way in which one man's reason issues orders to his own desires does not differ from the way in which it issues orders to the desires of other people. It is my reasonableness that, according to Plato, gives me the right—and the ability, which in this case is the same thing—to regulate both my own life and other people's.

I have spoken hitherto as if political action were the imposition of order or regularity upon something in itself innocent of these. This is, of course, not the case. It would be a mere fairy-tale to suppose that, first, there was a free, unregulated, non-political activity going forward, and that subsequently the political spirit arose from nowhere in particular (perhaps, as in certain romantic poets, from a race of serpents in this Paradise, called Tyrants) and killed its first fine careless rapture by riveting upon it the chains of law. Whatever fairy-tales may say, action that is innocent of all regulation is not action. The imposition of order upon pre-existing action can only mean the supersession of one type of order by another, which implies the cessation of one kind of action and the beginning of a different kind.

This conception will cause little difficulty to anyone who knows anything about primitive man. Before the rise of modern anthropological science, people might be pardoned for fancying that primitive man lives as he pleases at the dictates of instinct or some law of nature. We now know that primitive man lives under a system of primitive law, which he obeys in a blind, unquestioning way, not because he has explored and rejected all possible alternatives, but merely because he lacks the independence of thought which might enable him to think of another. For him, the alternatives are to enforce the traditional system, or, failing to enforce it, to expose himself to consequences all the more dreaded for being little understood. His political life is chiefly conceived as the maintenance of an existing order in actions; but this is only achieved by repressing every attempt to introduce alternative orders, whether these attempts come from an excess of instinct on the one hand or from an excess of intelligence on the other.

The progress of a people from a primitive to an advanced political condition, therefore, is not the imposition of order on what was once orderless. It is the substitution of one order for another; and (so far is it from being true that evolution introduces heterogeneity and complication into a previously homogeneous world) a civilized political system, like a civilized grammar, is often far less complicated than a barbaric. What, then, is the criterion of political progress? On what principle do people adopt a new political system as better than one already in existence?

The false answers to this question are many and easy. A political reform comes about, one might be tempted to say, for economic reasons, for moral reasons, for religious reasons, and so forth. But in fact it can only come about for political reasons. When the political spirit of a society is no longer satisfied with its existing structure, no longer finds that structure to express its own political aspirations, it alters it. And this process is really going on at all times. To speak of a stable political system as 'existing', and then as suddenly altered with a jerk by a so-called reformation or revolution, is to be deceived by appearances. Every fresh political action is in reality a modification of the whole political poise and attitude ('constitution') of the agent.

The agent is always a human being. This must be kept in mind if we are to avoid the mythology on which most political theories make shipwreck. We speak of a society, but the society is not anything except the people in it. Its actions are their actions. This will become clear by considering an instance. If we take the life of a single household, this life is political so far as it displays order, or regularity, or discipline. This discipline cannot exist except in the acts of the persons forming the household. It may partially show itself in a giving of commands by the father and an obedience by the rest[2]; but to say this is not to say that those who obey have no part in the family's political life. Their obedience is itself a participation in that life. Obedience is not the unintelligent submission to force; it is the understanding what another's purpose is, and the making that purpose one's own. Only a free agent can obey; only one who is conscious of himself as a source of independent political energy can recognize that energy in another, and co-operate with it. Giving commands to anyone who is not intelligently free is as useless as making faces at a blind man.[3]

The political life of a community is thus based on the freedom, the intelligent responsibility, of every member of that community. Freedom in this sense is quite compatible with being an infant or a slave in the eyes of the law; and, however true it is that some people are the better for being ordered about (which is certainly true of children, and, in Aristotle's opinion, of 'natural slaves'), it remains true that to order them about is to appeal to the freedom and the reason that are in them. And this is the reason why we order each other about, and are not content with ordering each man his own private affairs. Apart from the fact that a man's 'private' affairs are a metaphysical abstraction, the reason why all political life (that is, regulating and regulated life) is the life of a community is that the

---

[2] Though it must never be forgotten that only a small part of the community's political life consists in giving and obeying orders. The regularity or discipline of that life is partially achieved in this way; but far more by an unbidden acceptance of common rules which grow up nobody quite knows how. This is just as true of a family as it is (notoriously) of the law and custom of the British Constitution.

[3] Perhaps I may be allowed to illustrate this point by asking, why is domestic service such an unsatisfactory institution? Broadly speaking, because a person who is intelligent enough to be a 'good' servant in one sense (i.e. to do the work skilfully and to avoid waste and dirt) is too intelligent to be a 'good' servant in another sense (i.e. to be content with the social and economic status of service). Girls dislike going into service because they feel that they are not hiring their labour but hiring *themselves*: they prefer to work in a shop or a mill because here they feel that they are not (or not as much) giving up their freedom. It is generally recognized that there is a connection between the spread of education and the decline of domestic service: if you want servants, you must breed a stupid class for the purpose, and if they waste your bread and coal, that is the price you pay for the status of master.

principles of order know of no distinction between man and man. If the principles which I accept in my own personal life are valid for me, they must be valid for everyone whose situation is in the relevant respects like my own.

Political activity therefore proceeds not from outside inwards—from the group to the individual—but from inside outwards, from the individual to his fellows. The centre of gravity of political life lies not in the group, but in the individual. It is because chaotic and disorganized action would not be action at all that I must organize my life in order to have a life. And this organization can never be a hard-and-fast system invented and constructed once for all, or borrowed like the shell of a hermit crab; it is perpetually changing, and this change is my political history. But I cannot organize my own life without any reference to the organization of other people's. They, too, have a political history of their own; and our several political lives can no more be lived in so many watertight compartments than can our scientific lives; and for the same reason: in both cases we are pursuing a rational process which, because it is rational, is communicable, and has a validity not confined to persons.

The result of this is the formation of a network of political relations which grows in extent and complexity as we scrutinize it. All my contact with other human beings, and to some extent even my contacts with animals,[4] reveal a political aspect and form new threads in the network. Every such contact is an occasion for certain legislative acts, certain decisions as to how I shall conduct myself and what conduct I shall demand of my partner. These decisions are very often modelled closely on previous decisions by myself or other people, but none the less they are genuine decisions, original political acts. One such complex of political acts I call my family life; one, my membership of the Aristotelian Society; one, my British citizenship, and so on. In some of these relations I may be accustomed mainly to giving orders, in others mainly to obeying

---

[4] 'To some extent', I say, because my relations with animals are so overwhelmingly lop-sided. Certainly animals may have rights (e.g. it is recognized within the family that the cat has a right to its food; every member of the family recognizes this, and even the cat itself does something very like claiming its rights), and their rights may even be recognized by the state, as in fact many are by such Acts of Parliament as protect the nesting of birds, forbid cruelty to animals, etc. But though I can order my dog about, and my dog appears to order me about, the extent to which this can be done is exceedingly small as compared with the relation between a man and even a small child.

them; but we have already seen that the distinction between giving orders and obeying them (though highly important for the lawyer) is of merely secondary interest to political theory. In primitive law, no one gives orders, no one legislates, at all. Certain wise men know what the law is, but when they state it they are not commanding. On a committee, when a resolution has been amended and discussed and passed, all members are bound by it, but it is not a command issued by anyone to anyone. The question [of] whether I am a free and responsible citizen depends on whether I actually share in the political life of the country, not on whether any appreciable part of that life was my private invention.

Every association is a political association, and it is a mere superstition to regard 'the state' as having the monopoly of political powers and functions.

But does not this commit us to the anarchy of an unbridled and lawless political pluralism?

It cannot; for we have defined political life as order, and we cannot be expected now to define it as chaos. The solution is very simple. These various associations have a political character only so far as they involve the regulation of their members' intercourse by agreed laws. But the same people who are related in one way as fellow-churchmen are related in another way as fellow-townsmen, in another as fellow-clubmen, and so on. It is within the personality of the individual man that his relations with the members of his church, his town, and his club meet, and are organized according to the laws which he makes, or accepts, for the conduct of his own life.

There is thus, within the personality of the individual, a court of ultimate appeal before which all particular claims must state their case. His special loyalties to this or that association are not final; they are subject to revision, and, sitting in this ultimate court, he has the power of deciding to break with his church, or leave his town, or resign from his club. But this court of ultimate appeal is in reality nothing but the legislative power of his political life itself, which logically involves the power of repeal. It is simply the active centre or focus of his political activity.

This is the reality about which philosophers are talking when they speak of the State. That power, in any community, which has the last word on the question [of] what shall be law, is no mere partial or finite association; it is, in the political sphere, omnipotent. One ought not to be frightened of the word; it merely means active. The absolute or omnipotent state is simply political life, regarded as a focus of pure activity: all else is mere claim and counter-claim, mere pleas before this court.

But where is this absolute state? On earth, certainly; yet not visible in the outward form of parliaments and kings. It is within every one of us; and yet it can never genuinely exist, even within us, except in so far as it expresses itself in outward action, in which we co-operate in a common political life. These outward organs of its life are related to it much as learned bodies are related to knowledge or churches to religion. Superstition may confuse the outward organ with the life within, but only fanaticism can forget that a healthy inward life must create outward organs for itself.

The external and therefore finite organs of political life which are called states are necessary to the political life of the communities which they represent. They are not on a level with other organizations; for whereas other organizations—the family, the church, the trade union, and so forth—have other primary objects and are political only secondarily and because they cannot help it, the outward state is, on the whole, a deliberately and primarily political organ. Because this is so, we are apt to forget how small a fraction of the community's political life passes through the state's hands, and how much even this fraction arrives there in a predigested form. But to those of us who have been brought up in the school of Maitland this is an old story. Historically, as Maitland has taught us, the modern state is an almost fortuitous collection of functions left over from other bodies, and performs these functions in ways dictated to it by what those other bodies have already done. That is true; yet it remains none the less true that the modern state, however it has arisen, has established itself as *par excellence* the political organ of society, and whatever non-political work it may do is done because of some relation to its political function.

Bearing this in mind, it is not hard to give a theoretical solution to the problem of sovereignty. Sovereignty is merely a name for political activity, and those who would banish sovereignty as an outworn fiction are really only trying to shirk the whole problem of politics. But sovereignty does not belong to any determinate organization. It belongs only to that political life which is shared by all human beings.

The family is a sovereign body in so far as it genuinely regulates its own affairs. It cannot nowadays execute its own criminals, as it could in ancient Rome; but that is not a diminution of its sovereignty. It merely means that the family has ceded that particular function to the state. Whether the initiative in this change came from the one side or the other, as we have seen, makes no difference; that is mere archaeology; the point is that, here and now, this division of powers is accepted by both sides. The family does not want to hang its own criminals, and the state does not want to fix the hour of Sunday dinner. Thus we have two communities, each organizing its activities so as to fit in with those of the other. Each is limited by the other, and is to that extent finite and not possessed of sovereignty; but each is willing to accept the limitation, so that it becomes a self-limitation and therefore no real loss of sovereignty or power; for power is only really lost when one is unable to do what one wants to do.

The plurality of associations is thus no barrier to the sovereignty of each and all, just as long as the relation between them is controlled by a principle to which all assent. In this agreement, the limitation of each is a limitation based on its own choice, and therefore each remains sovereign. But what happens when they fail to agree? Each, then, becomes subject to limitations which it does not accept; it finds itself irked, hampered, crippled by the activities of the other associations, and this is the state of war. War in this sense is sovereignty failing to establish itself; it is not, as some think, the apotheosis of the state, but its bankruptcy; it is the failure to act politically. And it should be observed that war—not in the special sense of war with explosives or poison gases, but in the general sense of political strife—is the evil that is inseparable from the political good.

But this evil can always be overcome. The certainty that a satisfactory adjustment *can* be made as between the state and the trade unions, or parliament and the church assembly, or the empire and the League of Nations, is absolute. If sovereignty were an attribute of a substantially existing entity, an appanage of the outward and visible state, it would be impossible, for it would imply that the state both remained itself—that is, sovereign—and sacrificed its sovereignty. But the real sovereignty is the sovereignty of the political spirit itself, in its constant envisagement of problems such as these and its constant and never quite unsuccessful attempt to solve them. Sovereignty belongs to the family not as maintaining its rights against the state, but only as agreeing in a distribution of functions between it and the state. Sovereignty belongs to the state not as issuing futile commands to recalcitrant subjects, but as formulating a law which these subjects can accept for their own.

From this point of view there is no more reason why the state should resist inclusion in a League of Nations without, than why it should suppress associations within. So far, the theory I am putting forward is in agreement with the pluralists. I also agree with the pluralists that all bodies whatever (and I would go further and say all individuals) are doing political work, which is therefore no monopoly of the state. But in two ways I find myself compelled to differ from the pluralist.

First I would point out that the plurality of associations cannot form a body politic unless they are unified into a single whole by agreeing upon a common policy. Short of that agreement, they are not so many independent political agents, as the pluralist thinks, but so many warring factions, whose mutual hostility only serves to show that none of them has risen to the level of political action. And here I feel bound to remind you that the work of mediating between the conflicting interests of classes and sections, and finding the diagonal of the parallelogram of political forces, is precisely the work of that special organ which goes by the name of state.

Secondly, when I find the pluralist suggesting, as he sometimes does suggest, that the work of the state may be done better by delegating it to other bodies that are in closer touch with the actualities of social life, I know where I am. I am listening to the fallacies of historical materialism. I am being regaled with the old story that nothing is real but what is economic. To that, I reply that I rejoice in the recognition of the reality of economic values, but deplore the tendency to let this recognition go to one's head. Economic values are real, but they are no more real than political values. I am ready to agree that the modern state is an imperfect embodiment of the political spirit: just as I am ready to agree that the Royal Academy is an imperfect embodiment of the artistic spirit. But I should not listen with much sympathy to anyone who said, 'You admit the imperfections of the Royal Academy; then let us scrap it and have an annual exhibition of British Art organized by the Bank of England.' A person who said that would be merely warning me not to trust his judgement in matters of art.

The economic life has its own end, prosperity: the political life has its own end, peace. Peace and prosperity may very well go together, but they are not the same thing. They are pursued by the same people, for we all pursue them both; but we pursue them in different ways and with different weapons. The weapon with which we pursue peace is the external, historical fact which we call the state; and I have tried to show that the state in this sense is a weapon, or, to vary the metaphor, an incarnation of political action; no more, and no less.

5
Politics [1929]

Action can find a reason for itself in more than one way. First, actions condition one another in so far as the doing of one act involves the not doing of something else. From this point of view, as we have seen, action determines itself positively by determining itself negatively, or gives up one end in order to achieve another; and this aspect reveals action in its character as economic action, and goodness as utility. Secondly, actions condition one another in so far as the doing of one act commits the agent not to refraining from something else, but to doing something else. Instead of entailing a renunciation, it entails a further positive action.

The way in which one act may commit the agent to another is this: if I do anything whatever, the result is that in some way my circumstances are changed, and the act brings me into a new situation. No doubt the change takes place in me, as well as in my circumstances; but that does not matter for our present purpose. Now, in these new circumstances I must act in a new way: the old act cannot merely be repeated, because it is not appropriate to the new surroundings. Hence the doing of the earlier act commits me to the doing, later on, of something qualitatively different.

But this difference cannot destroy a certain identity. Though the two acts are qualitatively different, they are acts of the same agent, conscious of his sameness; and the circumstances are no more merely different than the agent. On both sides there is an identity persisting through the differences. There must therefore be a similar identity persisting through the action. I am doing something different from what I was, but this something different is only the same old thing, modified to meet the modifications in myself and my surroundings.

This identity which persists through different actions is what we call in ordinary language a plan. A plan is the unity connecting a number of voluntary actions in such a way that each of them entails

Extract from 'The Moral Philosophy Lectures [1929]', Collingwood MS DEP 10, pp. 104–11.

and is entailed by the rest. The plan, in the abstract, apart from the actions in which it is embodied, is nothing but an abstraction; it lives in the various acts which it unites, and is only a way of saying that they are united. For this reason, although it is possible to a limited extent to think out one's plans in advance of acting them out, this is not essential to their existence as plans; and usually our plans come into consciousness in a piecemeal kind of way, we extemporize them in the course of executing them. This does not mean that we have no plans: it only means that we do not always think of them in abstraction from the particular acts which they unite. The way in which a plan exists is like the way in which the plot of a story or the thread of an argument exists; an extemporized story may have a plot, and an extempore argument a thread.

A plan connects into a single whole a number of different acts; not merely numerically distinct but qualitatively different. In all these acts we deal with peculiar situations by the adoption of peculiar measures; but the situations have an undercurrent of identity running through them, so far as they all arise out of the carrying out of our one plan, and the acts have an identity for this same reason. Hence we may say that so far as they come under one plan the acts are all similar acts, and the situations all similar situations. Now the principle of dealing with similar situations by similar measures is called a rule; and therefore a plan may be expressed as a rule or system of rules.

What we call a plan cannot be carried out without what we call rules. If my plan is to go and have a cup of coffee in the town, I must, in carrying out this plan, obey certain rules made for the occasion: I must avoid all shops except the ones where coffee is to be had, I must provide myself with money, and so forth. If my plan is of a more extensive kind, like working for an honours degree, the rules automatically become more complicated. I may think it necessary to read for several hours every day, to write an entirely original essay every week, or to work for a certain number of days every vacation.

Following out a plan or obeying rules may seem the opposite of freedom, for if we stand committed to a given act because of our rule, we are not free to do the opposite. But this is a mistake. We are always free to break the rule. But breaking rules merely to show we can do it is cutting off our nose to spite our face,[1] for the power

[1] At this point in the manuscript the text changes from Collingwood's hand-written script to typescript.

to follow out a plan is a real power, something involving more rationality and therefore more freedom than the simple power to do what we like at any given moment; and if we give up the greater for the sake of the less, we gain nothing.

A plan consists of acts organized into a coherent whole; but it must be remembered that no plan consists of a definite finite number of acts; and that any act, however trifling, is already a plan and contains in itself a diversity of lesser acts. However far one pushes the analysis of a plan into its component parts, one never reaches an atomic act which cannot be subdivided. And similarly, however far one goes in adding act to act and forming more and more complex plans, every plan that really is a plan has the simplicity and unity of a single act. The entire plan, however extensive it may be, is put into effect by a single and indivisible act of will. This follows from the general nature of mind, and is familiar in ordinary reflection. No one thinks that it is possible to say how many acts precisely are involved in such a thing as the establishment of the Principate by Augustus. Everyone sees that such a thing is one act, within which we may distinguish however many component parts we like. The whole act falls of itself into certain main elements or phases, and the distinction between these is not arbitrary; what is arbitrary is the point at which our subdivision is arrested. Action which is explained or determined in this way, by reference to a plan consisting of a single action differentiating itelf into parts, or various actions uniting into a whole, is being considered, as a logician would say, in terms of quantity. Logical quantity is the determination of judgements as universal and particular; the categories involved are those of unity and plurality, which, in relation to their synthesis totality, become the moments of whole and part. A plan is a whole, a unity, of which a single act is a part, one of the plurality which compose the unity. And since the logical moments unity and plurality are transcendental, every act must be a plan and every plan an act; there can be no plurality which is not a unity, and no unity which is not a plurality.

It is this conception of quantity in action which forms the philosophical root of politics. Politics means the organization of activities. Empirically, it is the name for the particular kind of work which is done by a particular kind of organization called the state; but the state may do other things besides politics, for instance, it may carry on trade or scientific research: and on the other hand a great deal of political work in the proper sense of the word is carried out by agents other than the state, for instance, by municipalities, churches, families, and so forth. What makes it right to connect politics with the state in a special way, is that the state exists in order to carry out political work—not all the political work there is, but some of it—and anything else it does is done because it somehow arises out of that primary function. The state is not a philosophical conception, but an empirical fact; and, like all empirical facts, it must exhibit in itself all the transcendental characteristics of all reality; and therefore the actual state, although it is primarily political, cannot keep clear of the economic, religious, aesthetic, scientific, and all such questions. Its success in dealing with these questions depends on the firmness with which it grasps its own essentially political character, and realizes that it can and must deal with them only so far as that character at once compels it and enables it to do so. Thus, the state does not dictate the creed of the church; all it does to the church is to insist that churchmen in following their own religion shall not break its laws; but this means that it is responsible for making laws which churchmen can obey without ceasing to be good churchmen, and this again implies that it has the right of telling churchmen that it demands the same co-operation from themselves. There must be a *modus vivendi* between church and state, and to find this *modus vivendi* is primarily the work of the state. Similarly, the state must have the power of supressing seditious or obscene publications, and cannot be expected to surrender this right merely because artists assure it that the publications in question have artistic merit. This fact brings it, in a sense, into conflict with the artists; and it is easy to blame the state for meddling with a business, namely literature, not its own; but it is meddling not with literature as literature, but with particular literary works as incidentally violating the order which it has to maintain in society. The only valid objection to such interference by the state with literature would be that it is not necessary for the maintenance of this order; and that is a political question, not a literary one.

But the state has no monopoly of political activity. Political action, being identical with action in general so far as this organizes itself, is a transcendental; the state is a merely empirical fact. Political action is therefore bound to overflow the limits of the state and to appear wherever there is action of any kind. Historically, the modern state has grown up by the accumulation of functions, primarily political, which have devolved upon it because they were not already provided for by other means. The administration of criminal law by the king's justices has grown out of the principle that the king, like everyone else, had to keep the peace in his own house; and because the sphere of the king's peace was peculiarly wide, it was necessary for him to develop a complicated system of machinery for the enforcement of it. But no development of criminal law can ever take away the right and the duty of every citizen to keep his own peace in the sense of doing all he can to protect the peaceful life of the persons living in his house. If I turn a noisy and threatening tramp off my doorstep, or sit in judgement on my children for scratching each other's faces, I am keeping my own peace; and in doing this I am not trenching on the sphere of the state, because the state has never undertaken to keep my garden free of tramps or teach my children good manners. These are political functions; but not functions which any state is likely to claim for its own. The rules of any corporation, the statutes of a company, the regulations of a club, the routine of a family, are all political facts, and no less political are the rules which a man makes for his own guidance, and revises from time to time as occasion demands.

Regulation or organization, which is the essence of political activity, consists in thinking out a scheme of action, whether for myself or for others, and putting this scheme into practice. There is no essential difference between the two cases in which the scheme is for my guidance and for that of others respectively. All government is self-government, in this sense at least, that the ruler, the person who makes the scheme, must be a member of the community for which it is made, and conversely the ruled, the people whose activities are determined by the scheme, must acquiesce in it to the extent of obeying it, which they cannot be compelled to do, and will not do except of their own free will.

Liberty in the political sense, like several other conceptions which we have already examined, is on the one hand a philosophical or transcendental conception, and, on the other, an empirical. Much confusion may be caused by failing to distinguish these two aspects of the term. For example, when people say that they have not, but wish to have, political liberty, they do not mean that they are being forced to obey laws against their will, for no one can possibly force anyone to obey a law which he chooses not to obey—an elementary truth to anyone who has grasped the idea of the freedom of the will; they mean that they lack certain empirical facilities for making their wishes known to the people who legislate, or for replacing these by other people if they think that this would lead to greater attention to their desires. Politics as an empirical science is concerned with the various ways in which this may be brought about. Empirically speaking, there are some governments which are self-governments and others which are not; philosophically speaking, all government is self-government, for the simple reason that no one can be governed except by his own will. It would be a mistake to use this philosophical conception as an argument either for a so-called free or democratic constitution, on the ground that this alone corresponds with the idea of self-government, or against the proposal to introduce such a constitution, on the ground that the people who demand it already have all they can want. It would be equally mistaken to say that because, empirically, some peoples enjoy self-government and others do not, therefore the philosophical doctrine that all government is self-government is untrue to fact.

The first element in political action, the thinking out a scheme of action, is called in empirical politics legislation. The second element, the putting of this scheme into practice, is called execution, or administering the law. These two functions, however much it may be expedient to separate them in the practice of the state, are really indivisible. In administering the law, the executive and the judiciary must interpret it, and this means modifying it. In practice it is always found that the work of a legislative body and the work of a court of law overlap, so that the same body frequently does work of both kinds. The distinction between making rules and applying them is a purely ideal distinction, and the two cannot go on except in the closest contact. To forget this would mean falling into the error of supposing that a code of law was a rigid body of rules framed once for all and applying mechanically to all similar cases, without receiving any reciprocal determination from these cases in the process. To think of law in that way means thinking of the act of legislation as something wholly external to the subsequent act of applying or interpreting the law; but this is to hypostatize a universal, and to conceive the relation between the ground and the consequent as if it were the relation between two physical facts, each absolutely external to the other. The fact that we are speaking of mind and its activities ought to make this mistake impossible.

The judge who is interpreting the law is not bound by the act of the legislator as by something which requires no co-operation of his own; he is bound by it only as by something that is a part of his own present act of will. He chooses that there shall be this law which he is applying; not that he has the constitutional power to change it if he dislikes it, but that he accepts the act of the legislative body as helping him to decide the case he is called upon to settle. In accepting this law, and deciding to apply it to this case, he is reaffirming it as an act of his own will, just as certainly as I am reaffirming my own past decision to make a journey when I take the necessary ticket. The law as ground and the application of it as consequent must, as we have already seen, fall inside one and the same act, which is at once a choice of the universal, the law, and the particular, its application.

These same considerations apply to the rules devised by private persons for the guidance of their own activities. It is sometimes fancied that rules of conduct are unnecessary and pedantic, that the only good rule of conduct is to have no rules, and that the only effect of having rules is to tie your hands and make you unable to judge individual situations on their merits. This polemic against rules of conduct, however useful it may be as a warning against moral pedantry, is based on a confused idea of the nature of a ruler and is both self-contradictory and unpractical. It is self-contradictory, because to assert that you never make rules is to commit yourself to a rule, and a singularly sweeping one, which is none the less a rule for being framed negatively. It is unpractical, because no one would in practice give himself the trouble of asking himself, every morning, whether this was or was not a day on which he had better get up; whether he ought to eat anything today, whether it was his duty to start going about naked today, and so forth. The most determined theoretical opponent of rules does in fact observe them like other people. His opposition is based on a confused idea of what rules are, because he thinks of a rule as a dead mechanical thing which determines his acts in advance, before the data on which alone they can be justly determined are in his possession. But this is not the case; a rule has to be applied, and the application is a new act in which the rule is reaffirmed as relevant to this situation. It is idle to object to rules because one has made the mistake of supposing that the making and the application of a rule are two acts falling outside one another in the sense in which two events in the material world fall outside one another. Finally, it may be pointed out that when these people say that one ought not to be guided by rules but ought to judge each case 'on its merits', they do not observe that case[s] have 'merits' only if there are rules that apply to them. The merits of a case can only be expressed in terms of its relation to admitted or extemporized rules of conduct.

It is not the business of a moral philosophy to give a list of rules which the philosopher advises his disciples to follow, any more than it is the task of political philosophy to draw up civil and criminal, or even constitutional, codes. Any particular rule is an empirical fact, historically conditioned, and having its application only in the particular circumstances which compelled people to frame it. Codes of rules governing private conduct, or so-called moral codes, are in the same case. They belong to a certain historical environment, and do not apply outside that environment. This does not mean that they are useless. On the contrary, they are a necessary part of all practical life; though we must remember that their function belongs not to the moral activity in the special sense, but to the political activity. By framing and obeying such rules in the best possible way, we do not bring our lives into harmony with the idea of duty, but with the idea of orderliness, which is the governing conception of political action.

6
Politics [1933]

Parallel to the science of economics, which studies action under the special form of utility, there is a science which studies it under the special form of rightness or conformity to rule: namely politics. This is the second of the three moral sciences, economics, politics, and ethics in the special sense of that word, which deal respectively with utility, rightness, and duty. The peculiarity of politics lies entirely in the fact that it recognizes no form of goodness except conformity to rule. In practice, political work consists in thinking out rules and administering them, that is, procuring obedience to them; and in theory, political thought consists in grasping the conception of rule and all that it implies.

Like economics, politics as a science has an empirical part and a philosophical part. There are certain institutions, or organized complexes of human activity, which exist for the purpose of making and administering universal rules: these are generally called states, and on its empirical side politics may be described as the theory of the state. But the making and carrying out of rules is by no means an exclusive function of the state, nor yet its only function. Conformity to rule is a characteristic of all action whatever; there is no institution, whether state, church, school, club, or family, which does not make rules and impose them on its members, and there is no individual human agent who does not make rules and impose them on himself; and on the other hand the state, as a concrete historical institution, cannot confine itself to the making and administering of rules or laws, it must to some extent interest itself in economic and moral questions as well as political questions in the strict sense of the word. Consequently there cannot be a philosophical theory of the state, there can only be a descriptive, empirical, or historical account of it; the philosophical theory of politics can only be a theory of the political element which, though people are perhaps readier to discern its presence in the state than elsewhere, is not confined to it but is coextensive with human life.

Extract from 'Lectures on Moral Philosophy, 1933'. Collingwood MS DEP 8, pp. 99–103.

There is therefore a certain equivocation attaching to all the concepts of political science, according to whether they are understood empirically or philosophically. Empirically considered, the state is merely a certain type of historical phenomenon. [The following three sentences appear to have been inserted at a later date, perhaps late 1939.] It is a type of organization which has never confined itself to the strictly political function of making and enforcing rules, nor has it ever tried to monopolize that function as it exists in the lives of its citizens. It is therefore idle to attempt a deduction of the proper functions of the state from the principles of philosophical politics. In order to understand the state we must appeal not to philosophy but to history. [Here the insertion ends.] The state as we know it is an entirely modern thing, profoundly different in its essential character from, for example, the ancient Greek πόλις which from our point of view was at least as much a church as a state, being based on a common religion and regarding the maintenance of this religion as its *raison d'être*; nor is it to be expected that the state as we know it will survive indefinitely into the future. But philosophically the state is a mere name for the political function in human life; and in that sense it is an eternal necessity, for this function must be carried out in the consciousness that it is a function distinct from others. The same analysis will apply to the other concepts of politics. [Collingwood writes in pencil (insert war).]

As a philosophical science, politics is the science of rightness or conformity to rule. Its first task is to distinguish its own field from that of economics and that of ethics in the special sense, by distinguishing rightness from utility and duty. Rightness means the same as law and order, civil peace, and the various other phrases which the political consciousness has used as names for its own proper end. Its relation to utility is expressed by various phrases which define what the state does, as opposed to what the state is. The state is the rule of law; what it does is to provide for the security of person and property, the safeguarding of the subject's legitimate interests, and the like: that is to say, law and order or civil peace affords a framework within which interests or utilities may be pursued, and without which they could not be pursued without entering into a mutually destructive conflict. Its relation to duty is, roughly, expressed by such phrases as respecting the liberty of conscience: that is, laws do not usurp the function of conscience but provide a basis upon which the moral life of the individual may freely develop.

Having thus determined its relation to the other moral sciences by determining the relation of rightness to utility and duty, the philosophical science of politics must go on to study the concept of rightness in itself. This involves two terms: the universal, law itself, and the individual, the act of obeying law. These are both actions. Law is not something that exists independently of an agent; it exists only in the act of positing it or affirming it. How this fundamental act of positing law is related to the empirical conceptions of legislation, promulgation, codification, and so forth is a further question not belonging to the purely philosophical science, which only recognizes law as an act of universal will—not universal will in the sense of everybody's will, but universal will in the sense of a will willing what is universal, that is, willing not an individual act but an act of a certain kind. The individual act of obeying law is the recognition of the law as governing this individual case; this too is an action, namely the willing of the individual, or rather of the universal as present in the individual. The two acts of making and obeying law are thus the universal and individual aspects of one and the same act, and their unity in that act is its rightness.

The next step is to consider certain further implications of this concept. Because every law is universal, it is not enough to regard each individual act as right by reason of a special *ad hoc* law governing that individual case as a case *sui generis*. Such an *ad hoc* law must be a special application of wider laws, and the rightness of this individual act therefore implies the rightness of certain other acts. If this individual act is right here and now, it follows from the nature of rightness that similar acts, *mutatis mutandis*, are right— will be right and have been right—on similar occasions; and in the widest sense all acts and occasions have certain similarities to all others. Thus a single law involves an infinite complex of laws all systematically related, and the rightness of a single act involves the rightness of an infinite number of others; and so we arrive at the idea of law as a systematic whole, governing the conduct of life as a systematic whole, and forming in itself the detailed and articulated content of a single ultimate and purely formal law, the law that there shall be laws.

The task of practical politics is the realization of this idea. The idea itself is logically implied in the very conception of rightness, and therefore the task is an inevitable and universal one; the necessity of attempting it arises not only in the field of empirical politics, but wherever rightness is recognized and pursued as a form of value; that is to say, just as much in the self-government of a single individual's life as in the life of a state.

There are certain obstacles or difficulties in the way of this task, arising out of the nature of the task itself. The actions which we do and the *ad hoc* rules by which they are justified are a manifold of distinct actions and distinct rules; each must be distinct and independent of the rest, for if any were a mere consequence of another it would not be a genuine action. But this mutual independence, though not in itself a conflict, involves a possibility of conflict. It is always an open question how far the wider and more general rules implied in the *ad hoc* rules governing two distinct individual actions are capable of being brought into harmony. When they conflict, each act is indeed subsumed under the rules governing the other, but is subsumed not as right but as wrong.

A wrong therefore presupposes two different rights: it is right in relation to its own rule, and there is another right with which it is in conflict. We found, in the same way, that the inexpedient was always one expediency incompatible with a greater expediency. Unlike expediency, rightness cannot be compared with rightness as more and less right; how therefore are we to decide which of two conflicting rights is objectively right and which objectively wrong? Or are we to say that each is equally right in itself or equally wrong by reason of its conflict with the other?

This second alternative comes neare[r] to being exemplified in the case of war between two states, each claiming to represent a complete system of law and thus a complete organization of human life, which come into conflict over a question upon which each takes a line of action following necessarily from its own conception of right. In such a case each thinks itself simply right and the other simply wrong; and an impartial spectator can only say that each *is* right from its own point of view. But even here this is not the whole truth. The two states are not in fact two complete and independent systems of life; they are partners in a common life, sharing in a tradition which is wider than each of them or both together; and before the court of this tradition one may be objectively right and the other objectively wrong.

This gives a clue to the solution of our problem. A man forging a cheque may think he has a right to do so, because he desperately needs the money, and the man whose name he is forging will never know and will not be harmed by the loss of so small a sum. This reason gives the rule by which he regards this individual act, or any act resembling it in essentials, as justified. He may have no objection to applying this rule universally and saying that anybody else would have a right to do the same in his place. But he knows that in presenting the cheque over the counter of the bank he had better keep these views to himself, because it is impossible that they should be shared by the cashier. He knows, that is to say, that the rules on which he is acting are not only different from, but actually in conflict with, the rules of the institution within which and by whose means he is trying to procure his ends, namely the institution of banking. He is therefore willing the existence of two conflicting systems of law at once, not one for himself and one for other people, or one for this act and one for other acts, but both coinciding in one and the same act: a system of rules forbidding forgery and a system of rules permitting forgery.

This constitutes the objective wrongness of his action. And this is the only thing that can constitute objective wrongness in any action: namely its having a reason which on analysis will be found to consist of two conflicting rules. So far as crime (that is, breach of positive law) is genuinely wrong, and not a mere revolt against a certain determinate social order, this is why it is wrong; and this may serve to distinguish genuine criminals from those would-be critics and reformers of the social order who are all too easily mistaken for criminals by the men responsible for maintaining it. Every act of a rational man is subjectively right, or justified in his own eyes by *some* rule; a wrong act is one whose rule conceals a conflict between two incompatible rules; and the task of government in repressing crime, so long as what it represses is crime and not criticism, is the task of repressing action which is irrational in this sense, that is, based on a confusion of thought as to what the rule is which it obeys.

This is the basis of the philosophical theory of punishment. Men are punished, so far as they are rightly punished, not for neglecting their duties and acting otherwise than as conscience bids them, but for disobeying the laws of their country; and not for disobeying these laws in particular, but for disobeying law as law; that is, they are not punished for making private laws of their own—we all do that—but for making self-contradictory laws of their own. So far as punishment fails to conform to this criterion, it is exercised not by right but only by might. But so far as it does conform to this criterion, it is an essential element in all right action; it is the affirmation of the ideal of acting according to rule, operating negatively in the repression of whatever action is not according to rule; for a self-contradictory rule is no rule.

7

Punishment and Forgiveness

1. . . . . Forgiveness and punishment are generally conceived as two alternative ways of treating a wrongdoer. We may punish any particular criminal, or we may forgive him; and the question always is, which is the right course of action. On the one hand, however, punishment seems to be not a conditional but an absolute duty; and to neglect it is definitely wrong. Justice in man consists at least in punishing the guilty, and the conception of a just God similarly emphasizes his righteous infliction of penalties upon those who break his laws. The very idea of punishment is not that it is sometimes right and sometimes wrong or indifferent, but that its infliction is an inexorable demand of duty.

On the other hand, forgiveness is presented as an equally vital duty for man and equally definite characteristic of God. This, again, is not conditional. The ideal of forgiveness is subject to no restrictions. The divine precept does not require us to forgive, say, seven times and then turn on the offender for reprisals. Forgiveness must be applied unequivocally to every offence alike.

Here, then, we have an absolute contradiction between two opposing ideals of conduct. And the result of applying the antithesis to the doctrine of atonement is equally fatal whichever horn of the dilemma is accepted. Either punishment is right and forgiveness wrong, or forgiveness is right and punishment wrong. If punishment is right, then the doctrine that God forgives our sins is illusory and immoral; it ascribes to God the weakness of a doting father who spares the rod and spoils the child. If punishment is wrong, then the conception of a punishing God is a mere barbarism of primitive theology, and atonement is no mystery, no divine grace, but simply the belated recognition by theology that its God is a moral being. Thus regarded, the Atonement becomes either a fallacy or a truism.

And it is common enough, in the abstract and hasty thought which in every age passes for modern, to find the conception of

Reprinted from *Religion and Philosophy* (Macmillan, London, 1916), pp. 169–180, with the kind permission of the publisher.

atonement dismissed in this way. But such thought generally breaks down in two different directions. In its cavalier treatment of a doctrine, it ignores the real weight of thought and experience that has gone to the development of the theory, or broadly condemns it as illusion and dreams; and secondly, it proceeds without sufficient speculative analysis of its own conceptions, with a confidence based in the last resort upon ignorance. The historian of thought will develop the first of these objections; our aim is to consider the second.

The dilemma which has been applied to theology must, of course, equally apply to moral or political philosophy. In order to observe it at work, we must see what results it produces there. Punishment and forgiveness are things we find in our own human society; and unless we are to make an end of theology, religion, and philosophy by asserting that there is no relation between the human and the divine, we must try to explain each by what we know of the other.

(*a*) The first solution of the dilemma, then, might be to maintain that punishment is an absolute duty and forgiveness positively wrong. We cannot escape the rigour of this conclusion by supposing forgiveness to be 'non-moral', for we cannot evade moral issues; the possibility of forgiveness only arises in cases where punishment is also an alternative, and if punishment is always right, then forgiveness must always be a crime.

Forgiveness, on this view, is a sentimental weakness, a mere neglect of the duty to punish. It is due to misguided partiality towards an offender; and instead of cancelling or wiping out his crime, endorses it by committing another. Now this is a view which might conceivably be held; and if consistently held would be difficult to refute, without such a further examination of the conceptions involved as we shall undertake later. At this stage we can only point out that it does not deserve the name of an ethical theory; because it emphasizes one fact in the moral consciousness and arbitrarily ignores others. The fact is that people do forgive, and feel that they are acting morally in so doing. They distinguish quite clearly in their own minds between forgiving a crime and sentimentally overlooking or condoning it. Now the theory does not merely ignore this fact, but it implicitly or even, if pressed, explicitly denies it. To a person who protested 'But I am convinced that it is a duty to forgive,' it would reply, 'Then you are wrong; it is a crime.' And if asked why it is a crime, the theory would explain, 'Because it is inconsistent with the duty to punish.' But the duty to punish rests on the same basis as the duty to forgive; it is a pronouncement of the moral consciousness. All the theory does is to assume quite uncritically that the moral consciousness is right in the one case and wrong in the other; whereas the reverse is equally possible. The two duties may be contradictory, but they rest on the same basis; and the argument which discredits one discredits the other too.

(*b*) The same difficulty applies to the other horn of the dilemma, according to which forgiveness is always right and punishment always wrong. Just as we cannot say that forgiveness is a crime because punishment is a duty, so we cannot say that punishment is a crime because forgiveness is a duty. But the theory of the immorality of punishment has been worked out rather more fully than is (I believe) the case with the theory of the immorality of forgiveness.

(i) Just as forgiveness was identified with sentimental condoning of an offence, so punishment has been equated with personal revenge. This view has been plausibly expressed in terms of evolution by the hypothesis that revenge for injuries has been gradually, in the progress of civilization, organized and centralized by state control; so that instead of a vendetta we nowadays have recourse to a lawsuit as our means of reprisal on those who have done us wrong. But such a statement overlooks the fact that punishment is not revenge in the simple and natural sense of that word. The difference is as plain as that between forgiveness and the neglect of the duty to punish. Revenge is a second crime which does nothing to mitigate the first; punishment is not a crime but something which we feel to be a duty. The 'state organization of revenge' really means the annihilation or supersession of revenge and the substitution for it of equitable punishment. And if we ask how this miracle has happened, the only answer is that people have come to see that revenge is wrong and so have given it up.

(ii) A less crude theory of punishment as merely selfish is the view which describes it as deterrent, as a means of self-preservation on the part of society. We are told that crime in general is detrimental to social well-being (or, according to more thorough going forms of the conception, what is found to be detrimental is arbitrarily called crime), and therefore society inflicts certain penalties on criminals in order to deter them and others from further anti-social acts. It is the function of 'justice' to determine what amount of terror is necessary in order to prevent the crime.

Punishment so explained is not moral. We punish not because it is a duty but because it preserves us against certain dangers. A person has done us an injury, and we maltreat him, not out of a spirit of revenge, far from it, but in order to frighten others who may wish to imitate him. The condemned criminal is regarded as a marauder nailed *in terrorem* to the barn-door. One feels inclined to ask how such a combination of cruelty and selfishness can possibly be justified in civilized societies; and if the theory is still possessed by a lingering desire to justify punishment, it will perhaps reply that the criminal has 'forfeited his right' to considerate treatment. Which means either that he has cut himself off from our society altogether (which he plainly has not) or that there is nothing wrong in being cruel to a criminal; which is monstrous. If society is trying to be moral at all, it has duties towards a criminal as much as towards any one else. It may deny the duties, and have its criminals eaten by wild beasts for its amusement, or tortured for its increased security; perhaps the former is the less revolting practice; but in either case society is demonstrating its own corruption.

The deterrent theory, then, must not be used as a justification, but only as an impeachment, of punishment. But even if punishment is, as the theory maintains, a purely selfish activity, it must still be justified in a sense; not by its rightness but by its success. The question therefore is whether as a matter of fact punishment does deter. Now a 'just' penalty, on this theory, is defined as one which is precisely sufficient to deter. If it does not deter, it is condemned as giving insufficient protection to society, and therefore unjust. Society will accordingly increase it, and this increase will continue till a balance is established and the crime is stamped out. Those crimes therefore happen oftenest whose statutable penalties are most in defect of this ideal balance. The fact that they happen proves that the penalty is inadequate. Therefore, if the deterrent view is correct, society must be anxious to increase these penalties. But we do not find that this is the case. If criminal statistics show an increase, we do not immediately increase the penalties. Still less do we go on increasing them further and further until the crime is no longer attractive. If we may argue from empirical evidence, such as the infliction of the death penalty for petty thefts, it is simply not the case that increased severity necessarily diminishes crime; and yet on the theory it ought to do so. On the contrary, it sometimes appears that higher penalties go with greater frequency. To reply to this that the frequency of crime is the cause, not the effect, of the greater severity, would be to confess the failure of punishment as deterrent; for, on that view, severity ought to be the *cause* of *infrequency*, not the *effect* of *frequency*. The plea would amount to a confession that we cannot, as is supposed, control the amount of crime by the degree of punishment.

Thus the view that punishment is a selfish act of society to secure its own safety against crime breaks down. Its plausibility depends on the truth that the severity of punishments is somehow commensurate with the badness of the crime; that there is a connection of degree between the two. If we ask how this equation is brought about, the theory disappears at once. In punishment we do not try to hurt a man as much as he has hurt us; or even as much as may induce him not to hurt us. The 'amount' of punishment is fixed by one standard only; what we suppose him to deserve. This is difficult to define exactly, and common practice represents only a very rough approximation to it; but it is that, not anything else, at which the approximation aims. And the conception of desert reintroduces into punishment the moral criterion which the theory tried to banish from it. To aim at giving a man the punishment he deserves implies that he does deserve it, and therefore that it is our duty to give it him.

(*c*) Both these escapes, therefore, have failed. We cannot say that either punishment or forgiveness is wrong, and thus vindicate the necessity of the other. Though contradictory they are both imperative. Nor can we make them apply to different cases; maintaining for instance that we should forgive the repentant and punish the obdurate. If we only forgive a man after he has repented, that is to say, put away his guilt and become good once more, the idea of forgiveness is a mockery. The very conception of forgiveness is that it should be our treatment of the guilty as guilty.

Nor can we escape by an abstraction distinguishing the sinner from the sin. We punish not the sin, but the sinner for his sin; and we forgive not the sinner distinguished from his sin, but identified with it and manifested in it. If we punish the sin, we must forgive the sin too: if we forgive the sinner, we must equally punish him.

2. This absolute contradiction between the two duties can only be soluble in one way. A contradiction of any kind is soluble either

by discovering one member of it to be false, an expedient which has already been tried, or by showing that the two are not really, as we had supposed, incompatible. This is true, whether the contradiction is between two judgements of fact or between two duties or so-called 'judgements of value'; for if it is axiomatic that two contradictory judgements cannot both be true, it is equally axiomatic that two incompatible courses of action cannot both be obligatory. This fact may be obscured by saying that on certain occasions we are faced with two alternatives of which each is a duty, but the question is, which is the greater duty? But the 'greater duty' is a phrase without meaning. In the supposed case the distinction is between this which we ought to do, and that which we ought not; the distinction between *ought* and *ought not* is not a matter of degree.

Granted, then, that in any given situation there can be only one duty, it follows necessarily that if of two actions each is really obligatory the two actions must be the same. We are therefore compelled to hold that punishment and forgiveness, so far from being incompatible duties, are really when properly understood identical. This may seem impossible; but as yet we have defined neither conception, and this we must now proceed to do.

(*a*) Punishment consists in the infliction of deserved suffering on an offender. But it is not yet clear what suffering is inflicted, and how it is fixed, beyond the bare fact that it must be deserved. If we ask, 'Why is that particular sort and amount of pain inflicted on this particular man?' the answer, 'That is what he deserves,' no doubt conveys the truth, but it does not fully explain it. It is not immediately clear without further thought that *this* must be the right punishment. Punishment is fixed not by a self-evident and inexplicable intuition, but by some motive or process of thought which we must try to analyse. The conception of desert proves that this motive is moral; and it remains to ask what is the moral attitude towards a crime or criminal.

If we take the case of a misdeed of our own and consider the attitude of our better moments towards it, we see that this attitude is one of condemnation. It is the act of a good will declaring its hostility to a bad one. This feeling of rejection, condemnation, or hostility is in fact the necessary attitude of all good wills towards all evil acts. The moral action of the person who punishes, therefore, consists primarily in this condemnation. Further, the condemnation, in our own case, is the act in and through which we effect our liberation or alienation from the evil, and our adherence to the good. If a person is in a state of sin, that he should feel hostility towards his own sin is necessary to his moral salvation; he cannot become good except by condemning his own crime. The condemnation of the crime is not the *means* to goodness; it *is* the manifestation of the new good will.

The condemnation of evil is the necessary manifestation of all good wills. If A has committed a crime, B, if he is a moral person, condemns it. And this condemnation he will express to A if he is in social relations with him; for social relations consist of sharing thoughts and activities so far as possible. If B is successful in communicating his condemnation to A, A will thereupon share it; for A's knowledge that B condemns him, apart from his agreement in the condemnation, is not really a case of communication. But if A shares the condemnation he substitutes in that act a good will for an evil. The process is now complete; A's sin, B's condemnation, B's expression to A of his feelings, A's conversion and repentance. This is the inevitable result of social relations between the two persons, granting that A's will is good and that the relations are maintained.

Now this self-expression of a good will towards a bad is, I think, what we mean by the duty of punishment. It is no doubt the case that we describe many things as punishment in which we can hardly recognize these features at all. But examination of such cases shows that precisely so far as these facts are not present, so far as the punishment does not express moral feelings, and does not aim in some degree at the self-conviction of the criminal—so far, we are inclined to doubt whether it is a duty at all, and not a convention, a farce, or a crime. We conclude, therefore, that punishment—the only punishment we can attribute to God or to a good man—is the expression to a criminal of the punisher's moral attitude towards him. Hence punishment is an absolute duty; since not to feel that attitude would be to share his crime, and not to express it would be a denial of social relations, an act of hypocrisy.

(*b*) The pain inflicted on the criminal, then, is not the pain of evil consequences, recoiling from his action in the course of nature or by the design of God or man upon his own head; still less is it the mere regret for having done something which involves himself or others in such consequences. These things are not punishment at all, and ought never to be confused with it, though they may well be incidental to it. The pain of punishment is simply the pain of self-condemnation or moral repentance; the renunciation of one aim and the turning of the will to another. That is what we try to inflict upon him; and any other, incidental pains are merely the means by which we express to him our attitude and will. But why, it may be asked, should these incidental pains be necessary? Why should they be the only means of communicating such feelings? The answer is that they are not. The most perfect punishments involve no 'incidental' pains at all. The condemnation is expressed simply and quietly in words, and goes straight home. The punishment consists in expression of condemnation and that alone; and to punish with a word instead of a blow is still punishment. It is, perhaps, a better and more civilized form of punishment; it indicates a higher degree of intelligence and a more delicate social organization. If a criminal is extremely coarsened and brutalized, we have to express our feelings in a crude way by cutting him off from the privileges of a society to whose moral aims he has shown himself hostile; but if we are punishing a child, the tongue is a much more efficient weapon than the stick.

Nor does the refinement of the penalty end there. It is possible to punish without the word of rebuke; to punish by saying nothing at all, or by an act of kindness. Here again, we cannot refuse the name of punishment because no 'physical suffering' is inflicted. The expression of moral feelings, or the attitude of the good will to the bad, may take any form which the wrongdoer can understand. In fact, it is possible to hold that we often use 'strong measures' when a word or a kind action would do just as well, or better. 'If thine enemy hunger, feed him; for in so doing thou shalt heap coals of fire on his head.'[1] Sentimentalists have recoiled in horror from such a refinement of brutality, not realizing that to heap coals of fire, the fires of repentance, upon the head of the wrongdoer is the desire of all who wish to save his soul, not to perpetuate and endorse his crime.

But at this stage of the conception we should find it hard to discriminate between punishment and forgiveness. If punishment is to express condemnation, it must be the condemnation of a bad will by a good one. That is to say, it is the self-expression of a good will,

[1] 'If thine enemy be hungry, give him bread to eat; and if he be thirsty, give him water to drink. For thou shalt heap coals of fire upon his head, and the Lord shall reward thee.' Prov. 25: 21–2.

and that good will is expressed as truly in the act of kindness as in the block and gallows. But if the punisher's will really is good, he continues, however severe his measures, to wish for the welfare and regeneration of the criminal. He punishes him not wholly with a view to 'his good', because the punishment is not consciously undertaken as a means to an end, but as the spontaneous expression of a moral will; yet the aim of that will is not the criminal's mutilation or suffering as such but the awakening of his moral consciousness. And to treat the criminal as a fellow-man capable of reformation, to feel still one's social relation and duty towards him, is surely the attitude which we call forgiveness.

If forgiveness means remission of the penalty, it is impossible to a moral will. For the penalty is simply the judgement; it is the expression of the moral will's own nature. If forgiveness means the remission of the more violent forms of self-expression on the part of the good will, then such restraint is not only still punishment but may be the acute and effective form of it. But if forgiveness means— as it properly does—the wise and patient care for the criminal's welfare, for his regeneration and recovery into the life of a good society, then there is no distinction whatever between forgiveness and punishment.

(c) Punishment and forgiveness are thus not only compatible but identical; each is a name for the one and only right attitude of a good will towards a man of evil will. The details of the self-expression vary according to circumstances; and when we ask, 'Shall we punish this man or forgive him?' we are really considering whether we shall use this or that method of expressing what is in either case equally punishment and forgiveness. The only important distinction we make between the two words is this: they refer to the same attitude of mind, but they serve to distinguish it from different ways of erring. When we describe an attitude as one of forgiveness, we mean to distinguish it as right, from that brutality or unintelligent severity (punishment falsely so called) which inflicts pain either in mere wantonness or without considering the possibility of a milder expression. When we call it punishment, we distinguish it as right from that weakness or sentimentality (forgiveness falsely so called) which by shrinking from the infliction of pain amounts to condonation of the original offence.

8
Punishment

[The] conception of political action as action in general from the point of view of logical quantity, as the sythesis of universal and particular, or rules and their applications, serves to explain a number of problems relating to political philosophy—all these problems, in fact, so far as they really belong to political philosophy and not to the empirical science of politics. As an example, we may here take the conception of punishment.

According to the ordinary textbook of ethics, punishment is either retributive, deterrent or reformatory. It is retributive, if people are punished simply because they deserve to suffer for what they have done. It is deterrent, if they are punished as an example to others to warn them not to go and do likewise. It is reformatory, if they are punished in order that they may learn to amend their ways.

These theories of punishment may be regarded either as rival definitions of its essential nature, or as accounts of elements which co-exist in it. When they are regarded as rival definitions, it is easy to show that the only one which is satisfactory is the retributive theory. The deterrent theory is open to the objection that unless a malefactor deserves to suffer, it is immoral to inflict suffering upon him for the sake of frightening other people; and hardly less immoral for these others to allow themselves to be frightened by the sufferings wrongfully inflicted on the person whose acts they propose to imitate. It is only if the suffering of the malefactor is deserved, that is to say, if the retributive theory is correct, that deterrence is justified.

The same considerations apply to the reformatory theory. If punishment is an attempt to reform a man, that is, to make him change his habits, by hurting him, it is an immoral method of education unless his habits are such that he deserves to be hurt; and the victim ought to resist attempts to terrorize him into such changes, unless the methods used are such as he deserves to have employed upon him.

Extract from 'Moral Philosophy Lectures [1929]', Collingwood MS DEP 10, pp. 112–23

The common objections to the retributive theory are based on two misunderstandings. In the first place, it is thought that retribution implies anger. It implies nothing of the sort. To hurt a person because he deserves to be hurt is not the same thing as hurting him because you are angry with him. On the contrary, until you have stopped being angry with him you are no fit person to settle what he deserves. The assessment of desert is an act of cool judgement, not an expression of passion. Secondly, it is thought that the retributive theory leads to cruelty, because it implies the infliction of pain not for any ulterior reason but for its own sake. But the opposite is the truth. The deterrent and the reformatory theories do certainly lead to cruelty, because if one believes them one is logically bound to go on punishing people more and more harshly whenever one finds them doing what they ought not, on the ground that they are evidently not yet enough reformed or deterred. This tendency to a constant increase of penalties can only be checked by appeal to the retributive theory, namely by appeal to the principle that this crime does not deserve such violent punishment. It is noteworthy that when people argue for the increase of the penalty for a given crime, they generally rely on the deterrent or the reformatory theory; they say, either that people must be taught not to do this sort of thing, or that prospective criminals must be terrified by the severity of the statutory or customary punishment. The question 'Do they really deserve such a punishment?' represents the point of view not of excessive severity but of moderation.

There is a more serious objection to the retributive view. It is, that we are not in a position to find out what people deserve, because that is a delicate question of moral motives which can never be answered with confidence or accuracy. This is not altogether true; in point of fact, we can and do assess the moral responsibility of ourselves and others every day of our lives. Our assessments may not be infallible, but they have as good a chance of being fairly correct as most of the opinions on which we act. On the other hand, the objection gathers weight if put in this form: that a court of law is not the best place for forming such judgements, and that in reality courts of law, which punish malefactors, do not even attempt to assess their purely moral guilt. Perhaps a judge might be able to make up his mind both definitely and justly, before delivering sentence, as to the prisoner's moral condition; but he does not try to do so. The task which he undertakes is of a quite different kind.

Punishment is certainly retributive, whatever else it may be. No doubt, it always is other things as well; for whenever we punish anyone we are doing a concrete action which has all sorts of qualities; one of these and only one is its nature as punishment. When people insist that punishment is, or rather perhaps ought to be, deterrent or reformatory, what they probably mean is that it ought to be these things in addition to being punishment. Certainly a punishment is normally, perhaps necessarily, a public enough thing to affect the acts of others beside the person punished, and to bring to their notice the dangers of law-breaking. Certainly also, it must form an incident in the education of the sufferer, and ought to leave him more convinced than he was of the power and majesty of the law. In these ways it is unlikely that any good punishment could fail to produce good effects from the points of view of the reformatory and deterrent theories. We might even grant that if it did fail of producing such effects it would, other things being equal, be the worse as a punishment, because the failure to produce these effects would almost certainly be due to some lack of justice in the punishment itself. If the steps we take to punish people do not improve them and raise the general level of law-abidingness in the community, the probable inference is that our attempts to discover just punishments are being unsuccessful. It seems likely, then, that the deterrent and reformatory characters of punishment, though they are certainly not its essence, may be properties in the logical sense of the term, that is, attributes following necessarily from the essence. But we must return to the question [of] what this essence is.

We have seen that the objection to the retributive view is that when we punish we do not in fact necessarily try to ascertain the moral guilt of the criminal. What we do try to ascertain is whether he has broken the law, and if so, what law. The question is a legal question, not a moral question. Punishment is therefore a political action, not a moral action in the narrow sense of the term. In the wider sense, it belongs to the sphere of morals, for it belongs to the sphere of action, and indeed to that of voluntary and responsible action; but the proximate form to which it belongs is the political.

In order to make this clear, we might be disposed to look for cases in which we are certain that a person is morally guilty of something that he morally ought not to do, but equally certain that it is not the business of any law to punish him. Following this line of inquiry, it would be easy to find cases of admittedly immoral actions which are not, and in our opinion ought not to be, punishable by criminal justice. Moral delinquencies like greed, or laziness, or ill temper not only are not punished by the law of any land, but no one thinks they ought to be. And even people who have no doubt that it is wrong to get drunk do not therefore think that it ought to be made a criminal offence. The case of adultery is similar. Most respectable and morally sensitive people think that as a rule it is at least as wrong as many things for which people are rightly punished; but in English law at any rate it is not a crime. Historically, this is no doubt because it was once dealt with by the ecclesiastical courts; but when these courts lost their jurisdiction, the point is that cases of adultery were not brought within the scope of the king's courts. This seems to be a good case of an admittedly immoral action which is not punishable, and most people would agree in not wishing that the law should be so altered as to make it punishable.

But in thinking of cases like this, we must bear in mind that the courts are only a particular empirical means of inflicting certain kinds of punishments for certain kinds of offences; not the only means of inflicting any punishment for any offence whatever. A person who is not punished by a court of criminal justice does not therefore escape punishment altogether. He may for instance be punished by his neighbours refusing to call on him. Punishment, being a determination of action as such, is transcendental, and not limited in its occurrence to the activity of any particular agent or body. Any penalty for the breach of any rule is a punishment, and if a given act is not punishable in the courts, the reason for this may be that it is punishable elsewhere. The fact that adultery is not punishable by English criminal law is capable of being explained in this way; and the same is true of any similar case. Therefore we cannot argue that punishment turns on issues other than moral ones because there are cases of immoral action which meet with no punishment. The answer is that they are punished elsewhere than in the courts.

One result, however, we have already reached: namely that since it has been shown that some moral offences are not dealt with by the criminal courts, the purpose of criminal justice as administered in these courts is not the punishment of moral guilt as such. The law does not try to repress wrongdoing as such and in general, but only those particular kinds of wrongdoing which come within its province. The law tries to do only what it can do, and does not claim unlimited jurisdiction over the whole moral sphere.

But in saying this we are not advancing very far. There may be two reasons why the sphere of the law is limited in this way. First, it may be because criminal and immoral offences are different things, even though they may overlap; secondly, it may be because the law is an empirical fact which cannot deal with more than an arbitrarily limited portion of any field. The right answer is the second. It is therefore open to us to maintain that the purpose of punishment is not indeed the repression of moral guilt at large, but the repression of those kinds of moral guilt which can be repressed by the operation of such means as these.

What kinds of moral guilt are these? If we look carefully at this question, the only answer that we can give, I think, is 'none'. Moral guilt as such slips between the fingers of the law-courts. What they can do is to insist that the law shall be obeyed; not law in general, not even the laws by which a particular community lives, but those of these laws which are selected for this particular kind of treatment. Some of the laws by which a community rules its common life can be enforced by public opinion, without the help of the courts; some are so liable to change that if the courts tried to enforce them the rigidity which this would bring into them would defeat their own object. For instance, the length of the female skirt is regulated by law, but the law in question is in a process of constant and slight change, and it would be very difficult indeed for a court to enforce it. Women pay the most careful attention to the problem of ascertaining the present state of the law on matters like this, and obey it with the most exemplary loyalty. They even practise a highly efficient system of punishment which is brought to bear on those of their number who fail to obey it. The most usual methods of punishment are a kind of imprisonment, that is to say, deprivation of the right of free intercourse with the rest of society, which here takes the form of withholding invitations and visits; and a kind of reprimand or public expression of disapproval, which takes the form of gossip or backbiting. To describe those who administer these penalties as actuated by spite or personal enmity is foolish, and merely shows that one is not capable of analysing the logical structure of what is called fashion, which is one of the empirical ways in which the political spirit, the spirit of action organizing itself into systematic wholes, finds expression.

Now when a women refuses to call on another woman because she dresses in what is technically called an impossible manner, she is not making a moral judgement. She imputes no blame, in the moral sense of the word, to the offender. Perhaps, poor thing, she can't help being impossible; one doesn't blame her personally for it; but she is impossible, and must be treated as such. Similarly a judge may say that he does not know whether the criminal before him in the dock is a criminal by his own personal fault or not; he is not committed to any theory of the freedom of moral choice in persons brought up in an atmosphere of crime; all that concerns him is that the man, for whatever reason, is a criminal and must be treated as such. Here it is clear that in practice the attitude of administering punishment can be and is separated from the attitude of making moral judgements. This separation is no more than a convenient abstraction; as a matter of fact, since every action has more in it than its merely political aspect, no action is devoid of moral determinations, and in the last resort we only understand what an action really is if we refuse to make these abstractions about it and consider it in its totality. The judge would be a better man, and the law he administers would be the expression of the will of a better community, if he took the view that the criminal's moral state was relevant to the way in which he was to be treated. But though he would be a better man, he would not be a better judge; and though the community whose laws tried to take moral considerations into account would be a better community, its laws would not be better laws. The purpose of law as law, and of the judge as judge, is fulfilled when they have solved the purely legal problem of making and enforcing law. In the same way, a society which refuses to call on a woman merely because she dresses eccentrically is a worse society than one in which these considerations can be overruled by taking into account her reasons for doing so and her compensating merits; but the point is that it is not a more fashionable society. Fashion as such knows nothing of any question except whether its laws are obeyed.

The reason for which punishment is inflicted, then, is that law has been broken. But why does the breach of law require the infliction of punishment?

The answer depends on the principle that law, in the political sense, is not a name for a fact but a name for an act. When we speak of the laws of nature, we are referring to certain matters of fact; we mean that there are certain ways in which natural events happen. The laws of nature cannot be broken. There is no such thing as an offence against the laws of nature, what goes by that name in common language is an event in which our desires and interests are frustrated by the operation of natural law, as when we eat too much and get indigestion. There is no 'law of nature' against overeating, but there is a 'law of nature' that if you overeat you will be upset. Overeating and being upset is not a breach of this law, but an instance of it.

A law in the political sense, which of course is the original and literal sense, is an act on the part of a legislator. It is the choice of a certain plan of action. In so far as the legislator is an individual person, the plan is a plan for his individual actions; but even in this case it can be carried out only if other people do not effectively oppose it, but co-operate with him at least to the extent of not getting in his way. If the legislator is a community, agreeing together upon a common plan of action, the plan is binding on all members of the community so far as they actually share in its common life. In order that the plan should be binding on any particular member of the community, it is not necessary that he should have had a voice in the origination of it. 'No taxation without representation' may be a very good principle for societies which have arrived at a certain stage of political education, but if it had been a primary principle of all politics, no one would ever have reached that stage. Civilization could never have begun. However desirable and admirable it may be that every adult member of a society should have a voice in its public affairs, it is not indispensable to the existence of a society that this should be so. And therefore the members of a society have an obligation to obey its laws which is not derived from their having helped to make them. This obligation is really derived from the fact that there must be laws, because action must, of its own nature, organize itself quantitatively into wholes in which the general plan determines the particular details. Some laws there must be, and therefore some way of determining what they are to be is indispensable. The obligation to obey law in general is the result of this. However much anyone may rebel against particular laws and particular rules, he cannot rebel against law or rulers as such. Even the extremest theoretical anarchist does not rebel against law, he only rebels against particular ways of making and enforcing laws, on the explicit ground that law can be made and enforced better without them.

But granted that the obligation to obey law in general is absolute and cannot be escaped, what is the nature of the obligation to obey any particular law? Clearly, we cannot argue that because law in general is inviolable, therefore any particular law is inviolable. A particular law is a particular plan for the organization of the activities of a society. As long as this plan is a plan at all, it is someone's plan. Someone is promoting or willing it. To disobey the particular law, therefore, is to put oneself in conflict with the will of this person. This conflict of the wills is what is called a breach of the law.

But a breach of law is more than a mere conflict of wills. The strife between political parties is a conflict of wills, but it is not a breach of law, unless it is so carried on as to involve breaches of the constitution. This is because the conflict between political parties is a difference as to what the law ought to be. So far as this is still an open question, there can be no breach of law. No one can break a law that has not yet been made. A breach of law implies that the law in question exists, which means two things: first, that the legislator or sovereign has willed the law, secondly that the subjects have accepted it. In the absence of this second condition, the law is what is called a dead letter, that is, a law which, although it is on the statute book, is not enforced.

The criminal, then, is a person who sets his own will in opposition not merely to the will of the ruler or sovereign, in which case he would be not a criminal but a rebel, and perhaps even a hero, but in opposition to the will of his fellow-subjects in so far as they are in the habit of obeying this particular law and are thereby making it something more than a dead letter. The general body of a society is agreed upon a certain plan of action, and someone sets himself in opposition to this plan. He finds himself struggling against the combined wills of all his fellow-citizens; the whole weight of the corporate will of his society is against him; and his inability to overcome this mass of opposition is the essence of his punishment. It must always be painful, simply because it is a failure, a passivity, the breakdown of an attempted action; and this, as we have already seen, is of the very nature of pain.

It is not open to the society to refrain from punishing such a person. If the society which is resolved to organize its activities in a certain way finds that one of its own members will not conform to this organization, it cannot simply for that reason change its mind and reorganize itself in a way to which he will not object, because there can be no guarantee that such a way can be found. If one man objects to one law and would prefer another, there is no way of being sure that this other, if it were adopted, would encounter no similar opposition. The fact of crime certainly supplies a motive for wishing to change the law, because it is always the law's fault that people cannot obey it; but it does not supply a motive for changing the law, because it affords no criterion by which a better law can be devised.

When a law is disobeyed, therefore, the society whose law it is will no doubt ask itself seriously whether a better law could not be found; but pending this discovery it cannot acquiesce in the breach of the existing law. To do so would be to surrender the plan of action which at present is controlling its movements, without adopting any alternative plan; and this is impossible, because, as we know, all action must organize itself according to some plan or other. For the present there is no option but to override the will of the objector and carry on with the plan to which he objects. This is called punishing him and enforcing the law.

The essence of punishment is this frustration of the criminal's will. He is not allowed to do what he wants to do, because his choice runs counter to the choice of the society in which he lives. He is crushed because he is in the minority; it is a mere question of brute force, and the force can no more help crushing that which impotently opposes it, than it, in its turn, can help being crushed. But although we call it force, it is really nothing but will. It is only the determination of the individual members of society to live according to certain rules, that supplies the force by which the will of the dissentient is broken. And it is his own fault, in the sense that if he had a better plan for the organization of society he might propose it and might find people willing to accept it, but in so far as he is merely a criminal he has no plan, he merely has a disinclination to fall in with the plans accepted by others. If he says he has a better plan, he thereby becomes a politician, and imposes upon himself the duty of bringing it persuasively before the public eye. If he says that he does not see why there should be any of these nasty plans, and that in his opinion everyone ought to be free to do as he likes, he is merely talking without thinking, for if he thought he would see that doing what one likes means carrying out one's plans, and that this is only possible if other people will fall in with the plans. Thus there is no possible defence for the criminal; his action is illogical in the sense that it ignores the necessary conditions of all activity. The criminal is trying to have his cake and eat it too; he is trying to break other people's rules of life, and to make other people obey his; he is, in fact, not simply denying the conception of law but affirming and denying it at once. His punishment, in the sense of the frustration of his will, is the logical consequence of this contradiction in his will. He cannot succeed, because what he is trying to do involves a self-contradiction; and his failure is his punishment.

Punishment is thus the automatic result of crime. In order to avoid punishment, a crime must become something other than a crime; for instance, a successful revolt against an authority whose pretence to power is a hollow sham. Here the so-called crime is not a crime, because the so-called law that is broken is not a law but a dead letter. Where law is really law, crime is punished not by the choice of any voluntary agent but by the necessary consequence of its criminality, that is, its conflict with the plan according to which society organizes its own activities.

Though punishment is always automatic, however, it is always an act of will. We cannot refrain from punishing, but our punishments are acts deliberately done. We must, whether we will or no, punish criminals, because the punishment is merely our continuance in the pursuit of our old and settled aims in spite of the attempted and unsuccessful frustration of them. But we can choose the way in which we shall punish. We can ask ourselves whether the criminal is capable of reformation, and if he is, we can punish him in such a way as to bring this about. We can ask whether the punishment would be best inflicted by the person immediately wronged, or by a third party. We may ask whether this or that existing body is best adapted to deal with crimes of this or that kind. And we always have to decide whether it is better to devise our punishments so as to make the criminal obey the law for the future, or so as to prevent him from interfering with other people's participation in the social system which the law expresses. These two elements, positive and negative, enforcing the law in the sense of making the criminal obey it, and enforcing it in the sense of preventing him from disobeying it, are the grounds on which the two main types of penalty rest. On the one hand, if we wish to bring the criminal within the system outside which he has put himself, we shall aim at reformatory penalties, and try to make punishment coincide with a special kind of education. On the other hand, if what we want is to remove a hindrance to other people's enjoyment of the rights recognized by law, we shall aim at putting him out of the way by imprisonment or death.

The relative emphasis on these two aspects of punishment is an index of the degree to which a society is confident of its own political force. If it fears that the shock of crime may disorganize it and make enforcement of its laws impossible, it will tend to refuse the benefit of the doubt to criminals, and assume that nothing can be done with them except get rid of them. If it has more confidence in itself and thinks that its laws are so reasonable that they are more likely to be obeyed than not, and that its people are sensible enough to be on the whole the friends of law and order, it will tend to treat all crime as a temporary aberration in the career of a person who, with proper treatment, can be brought back into the path of civic virtue. In that case, it will devise its punishments first and foremost as educational institutions; it will not cease to punish, but it will look forward beyond the punishment to the future life of the criminal as a reformed member of society.

This is not to say that capital punishment, with what may be called its minor degrees, imprisonment and penal servitude, are unnecessary. If they were not necessary they would not have been invented. But their necessity is not absolute, but relative to certain forms of political organization and certain degrees of political intelligence. Some societies certainly cannot get on without capital punishment, because they are politically backward and relatively unaccustomed to obeying laws. It is better to hang your murderers than to pretend you are civilized enough to do without hanging them if you really are not; but it is better still to be more civilized and to carry on the work of organizing society without the frightful waste that is caused by destroying an appreciable number of its members because you cannot invent a system into which they can fit without disaster.

9
Monks and Morals

Most of our crew had visited the monastery,[1] and none without being charmed and a little awed by what they found there: the atmosphere of earnest and cheerful devotion to a sacred calling, the dignity of the services and beauty of their music, the eager welcome and the loving hospitality, and above all the graces of character and mind which the life either generated in those who had adopted it or at least demanded of aspirants to it and thus focused, as it were, in the place where that life went on. They were all the more ready to talk about these impressions because they thought them a trifle paradoxical; they were the least little bit in the world ashamed of such feelings, because to have them seemed a kind of treason to their upbringing.

They had been taught that monks were at worst idle, self-indulgent, and corrupt; at best selfishly wrapped up in a wrong-headed endeavour to save their own souls by forsaking the world and cultivating a fugitive and cloistered virtue. They had, I suspect, been taught that the best was worse than the worst; for whereas a vicious monk was a sinner to be saved, and from another point of view a man doing his best, like most men, to have a good time, a virtuous monk was a man irremediably sunk in the deadliest of moral errors: a man who had renounced the primary duty of helping his fellow men, and had thus corrupted the best thing in human nature, the moral principle itself, into the worst, a purely individual and self-centred quest of salvation.

The music, for example. My friends were not utilitarian in any gross and barely material sense. They would have had nothing but praise for a man who should retire from the world in order to perfect himself in the performance or composition of music. But when that

Reprinted from *The First Mate's Log* (London and Oxford, OUP and Milford, 1940), 145–53, with the kind permission of Oxford University Press.

[1] The monastery to which Collingwood refers is that of the Monks of the Prophet Elijah situated near Pýrgos on the Greek island of Santorin (Thira). Collingwood visited the monastery in 1939 during a sea voyage on a schooner yacht crewed by Oxford students and himself.

had been done he ought to justify his retirement by coming out of it and making his performance or composition a blessing to his neighbours; not a means of filling their bellies, but a means of enriching and enlarging their minds. But what could be the social value of music, however beautiful, performed however beautifully to no audience?

These ideas, if unopposed, would have had no sense of paradox. They would merely have led to disapproval of persons who, as persons, were admittedly charming. The paradox arose because my friends were capable of making judgements not derived from these ideas, and in this case incompatible with them, and because they had the courage to abide by their judgements. They had seen these monks. They had lived for a time in close converse with them. They had found in them no discernible traces of the moral faults they had been taught to associate with the monkish profession. They had judged them to be good men, although by the standard of social utility they ought to have been bad men; and on reflection the judgement stood firm.

Finding myself drawn into discussions of this kind, especially of nights in the cockpit with Dick and Stephen, who now made up the Third Watch, I raised the question. How are you to judge social utility? Granted that social utility is your standard for the value of a man's life, how is the standard to be applied? Perhaps it would be reasonable to accept, at least provisionally, the judgement of those neighbours to whom, you say, the man should be useful, as to whether in fact they find him useful. To accept this judgement is to accept the principle that any man is useful to the society in which he lives in so far as his work fulfils a function which, in the opinion of that society, needs for its own welfare to be fulfilled.

Suppose a man devotes his life to the study of pure mathematics. Is he to be condemned for living on a selfish principle? Not, as my friends readily admitted, on the ground that pure mathematics cannot feed the hungry. Pure mathematics, apart from any consequences which may ultimately come of it, is pursued because it is thought worth pursuing for its own sake. In order to judge its social utility, then, you must judge it not by these consequences but as an end in itself.

What is more, you cannot judge the social utility of a pure mathematician by asking whether he publishes his results. Unless there is value in being a pure mathematician, there is no value in publishing works on pure mathematics; for the only positive result these works could have is to make more people into pure mathematicians; and a society which does not think it a good thing to have one pure mathematician among its members will hardly think it a good thing to have many.

The social justification of pure mathematics as a career in any given society, then, is the fact that the society in question thinks pure mathematics worth studying: decides that the work of studying pure mathematics is one of the things which it wants to go on, and delegates this function, as somehow necessary for its own intellectual welfare, to a man or group of men who will undertake it. A test for this opinion is that the society in question should be grateful to the pure mathematician for doing his job, and proud of him for being so clever as to be able to do it; not that every one else should rush in to share his life, but that even if his neighbours feel no call to share it they should honour him for living as he does. The fact that they do so honour him is a proof that they want a life of that kind to be lived among them, and feel its achievements as a benefit to themselves.

If this test is applied to the monks of the Prophet Elijah, the answer is not in doubt. Their neighbours are obviously and outspokenly proud of them. You had only to recollect the tone in which people you met asked you, 'Are you going to the monastery?' in order to realize that.

But there is an easy way of driving a coach and four through an argument of this kind. You have only to breathe the blessed word 'superstition'. The respectful neighbours are, of course, an ignorant, uneducated, unenlightened, superstitious folk; the monks are parasites who live on their ignorance and earn through their superstition a respect they do not deserve.

This is a criticism fatal to the line of thought I was suggesting, unless you are prepared to counter-attack it by asking what 'superstition' is and what part is played in human life by the things to which you give the name. But this would involve a good deal of difficult and perhaps humiliating thought. It is simpler to approach the question from another angle, thus:

The assumption underlying this appeal to the idea of 'superstition' is that the value of a certain man or group of men to the society in which they live cannot be decided by appeal to that society's own judgement, because on this particular question the persons who constitute the society, or whose judgement we refer to when we speak of the society's judgement, are bad judges, whose verdict is falsified by distorting influences of some kind or other, never mind what, which we call superstitions. But even if these cannot answer the question, the question still arises, and somebody must answer it. Who shall this be? Plainly the same person who, by using the word 'superstition', has exploded the claim of his rivals, the persons to whose principles he has applied the word.

The people of Santorin are proud of their monks, and this shows that, according to local opinion, the monks are a valuable section of society, whose work of prayer, and praise, and meditation is work that needs to be done, though not every one has a duty to do it. But, you say, that is only because the people of Santorin are so ignorant and superstitious that they cannot form a true judgement of what needs to be done and what does not. Very good: then it is for us, whose intellectual superiority entitles us to call them ignorant and superstitious, to use that same superiority for doing well what they have done badly. They have judged that their monks are doing a good work, and are of value to themselves. We are not content with their judgement because we think it has been made on wrong grounds. Let us make up our minds what the right grounds would be, and on those grounds make a judgement of our own.

What would the right grounds be? Social utility, you say. Nothing in the activity of one man or class of men is good unless it is useful, for its utility is what constitutes its goodness. I reply that this cannot be true, because it is self-contradictory. An action is useful because it leads to some other action. If this second action is desired only for its utility, that is as much as to say it is desired only because it will lead to a third action. Sooner or later, this series must end; an action must be reached which is desired not only for its utility but for its own sake: not only because it is expected to lead to something else, but because in itself it is regarded as good.

If utility is the only goodness, if nothing is good except in so far as it is useful, there is no utility and therefore no goodness: just as, if no commodities had any value except an exchange-value, none would have even exchange-value, because no exchange would be worth making. To judge all human action by the standard of utility is like establishing a paper currency in which notes can be exchanged only for other notes, never for gold and never for food or drink, tobacco, or railway-tickets, or the services of professional men. To have a currency of that kind is to be bankrupt; and the same name applies to having only a utility test for the value of human activities.

Sooner or later the judgement that something is good because it is useful rests on the judgement that something is good in itself, irrespectively of whether or not it is also useful. This shilling is some good to me because it is useful to me, and not for any other reason; and it is useful to me because by exchanging it I can get things that are good in themselves, things I desire for their own sakes: a fire when I am cold, a meal when I am hungry, or a lift in a bus when I am tired. Nothing is disqualified for being useful by being good in itself; a lift, which is good in itself as constituting a rest for weary legs, may also be useful as bringing a man nearer to his destination; but it is only because some things are good in themselves that anything can be useful.

The utilitarian trick of judging the worth of all human activities by assessing their utility is therefore logically nonsensical, and hence unworthy of any one who claims to be an educated and enlightened person; and it is morally disastrous, because it is the first step on the road to a moral bankruptcy brought about by some process in the moral life analogous to inflation in economic life. Inflation pushed to extremity means that real commodities, the things we really want to buy cannot be bought; all we can handle is stuff that is called money; but nobody wants money, people want the things that money can buy, and if money cannot buy things it forfeits the very name of money. So the moral bankruptcy of which I speak is the experience of finding that life is not worth living, because everything one does is done in the hopes of purchasing by its means a satisfaction which never comes. The way to avoid this moral bankruptcy is to stop judging the value of actions in terms of utility, and to judge them in terms of intrinsic worth.

Well then, let us try to do it. Here is Santorin, inhabited by such and such persons whom we have met, having such and such manners and customs which we have observed. There is something which we may call the Santorin way of life. It is not a mere aggregate of disconnected units; it is one pattern into which the monks and the children who gave us grapes and the girls who gave us water and the unknown person who painted the words of welcome at Pýrgos all fit as parts. What do we think of the Santorin way of life? Do we think it is a good way or a bad way? Of course we do not know very much about it; we have been in the island only four days, and our judgement will very likely be superficial. But we are not asked for a judgement which no further experience could correct. We are asked for a judgement based on the experience we have had: a judgement based on facing the facts that have come our way, facing them with an unprejudiced eye and an open mind.

And if we find ourselves concluding that the Santorin way of life is a good way, and that this commendation of Santorin life as a whole includes the monks who are part of it, we shall know what to do if we find ourselves adding: 'All the same as a good protestant I don't hold with monks; as a good secularist I don't think a life of prayer and meditation can be of any value; and as a good utilitarian I don't allow myself to praise anything except what I can praise for its utility.' We shall reply to ourselves: 'What is the use of travel if it doesn't broaden your mind? And how can it do that except by showing you the goodness of ways of life which, according to the prejudices you have learned at home, ought to be bad?'

In this way the traveller who began by thinking the men of Santorin ignorant, unenlightened, and superstitious may possibly, unless he is very careful, find within his mind a court sitting wherein the men of Santorin rise up in judgement against his own world and against the protestantism, and secularism, and utilitarianism of which it is so proud; and as judge in that court he may find himself obliged to take their part against his own world; so that if, later on, his own world should accuse him of not worshipping the -isms that it worships, and of corrupting its young men by imparting to them his heresy, he would have to admit that the accusation was just.

10

Duty

When we speak of an action or 'doing something' we commonly refer not to a completely individualized action but to a specified kind of action: not an individual, but a particular which may be individualized in various ways. We speak of returning a book, but 'returning a book' is not the name of an individual action, it is the name of a kind of action which may be done in many different ways. We speak of paying a penny, but paying a penny means paying any of the various pennies which one has in one's pocket. The rule which says that I must pay a penny says nothing about which penny I must pay. We speak of writing a certain letter, and say that it is one's duty to write it; but 'it' means only a letter of a certain kind, and the duty to write it means only the duty to write any letter of the kind specified: so that 'duty' in this case has its regularian sense, and the imperative one obeys in writing the letter is a disjunctive imperative. We are so much accustomed to using phrases like 'an action' or 'doing something' in this specific, non-individual, sense that it is difficult to use the same phrases to denote an individual action. If we wish to use them, we must take trouble to think what we mean: to make sure we are using them to denote not something that can be done in different ways, on different occasions, and so forth, but something in which there is no distinction between what is done and how it is done, no distinction between what is done and when it is done, and the like.

Because duty is always completely individualized, it involves a special and unique degree or kind of 'obligation'. Obligation in general is merely the denial of caprice. To be under obligation means to lack the particular kind of freedom which belongs to capricious action. Utility involves obligation: by choosing an end, one restricts one's freedom in so far as one commits oneself to acting in such a way that the action is means to that end. If I decide to save a tenth of my income, I thereby put myself under an obligation to

Extract from 'Goodness, Rightness, Utility: Lectures delivered in HT 1940'. Collingwood MS DEP 9, pp 69–77. 1940 lectures on Moral Philosophy.

spend only nine-tenths of it. But which part of it I spend and which part of it I save remains undecided. I am under no obligation to save any one part of it. I can still choose capriciously which part to save and which part to spend. Obedience to rule involves obligation; by choosing to recognize a certain rule, I restrict my freedom in so far as I commit myself to acting in such a way that one does nothing contrary to the rule. But there remain various possible ways of obeying it; and the obligation to obey the rule involves no obligation to adopt one of these rather than another. I can still choose capriciously which alternative I shall adopt. But the decision to do my duty is a decision to do one individual action for which no other can be substituted. Nothing is left to caprice.

Hence, of the three reasons for choice,[1] the third alone is a complete reason. Choice is always choice to do an individual action. Why do I do it? The answer 'because it is useful' explains only why I do an action leading to a certain end: not why, among the various possible actions which might have led to that end, I choose this and not another. The answer 'because it is right' explains only why I do an action of a certain kind, specified by the rule which I obey; not why, among the various possible actions conforming to that specification, I choose this and not another. But the answer 'because it is my duty' is a complete answer. What I do is an individual action; what it is my duty to do is an individual action; if what I do is my duty these two individual actions are one and the same. I do this and no other action because this and no other is the action it is my duty to do.

The relic of caprice which is still found in utility and rightness thus disappears in duty. The obligation to do my duty is an obligation involving every detail of what I am to do. Nothing is left to caprice; and for a person who means caprice when he says freedom, freedom has vanished. A person who does his duty has no option; he has got to do exactly what he does; he has no choice. The consciousness of this complete obligation—complete in the sense that it covers every detail of what is to be done and leaves no option anywhere—is a universal feature of duty. It may appear to involve the complete negation of freedom, but that is only because freedom is falsely identified with caprice. A man who knows that he has got to do exactly what he does, and that he has no option left

---

[1] Because it is useful; because it is right; and, because it is my duty.

anywhere, is in this state of complete obligation only because he is resolved to do his duty. That resolution is the act of his own will; and hence the apparent absence of freedom is not a genuine absence of freedom.

Capricious action is irrational action; and to say that there is no room for caprice in duty means that duty is completely rational. To say, 'I choose this because it is means to an end which I choose,' is to give a reason which is only a partial reason, for it explains not why I do this individual action but only why I do some action leading to that end; it affords no reason why among the many actions which might lead to that end, I choose to do this one. But this is only one of two unexplained elements in utilitarian action. Utilitarian action consists not only of willing the means, it consists also of willing the end; and the utilitarian analysis offers no reason, and in its very nature can offer no reason, why I will the end. This is why Henry Sidgwick said that utilitarianism could only provide a theory to explain our choice of means (he did not, I think, see that even of this it could offer only a partial explanation) and that utilitarianism in respect of means involved intuitionism in respect of ends. Intuitionism, as I have explained, is not a theory; it is only the denial of a theory; what Sidgwick was saying, therefore, was that the choice of ends is purely capricious.

To say 'I choose this because in choosing it I obey a rule,' is likewise to give only a partial reason. It explains not why I do this individual action but only why I do any of the possible individual actions between which the rule leaves me free to choose. And this choice between alternatives is only one of two elements which regularianism fails to explain. Regularian action consists not only in choosing to obey a rule, it consists also in recognizing the rule; and the regularian analysis can offer no reason why I recognize the rule. Here again, therefore, one is tempted to fall back on intuitionism, that is, on caprice.

This double element of caprice or irrationality, which is present both in utilitarian action and in regularian action, disappears in the case of an act done because it is my duty. For in the first place, doing this action because it is my duty leaves no freedom of caprice as to how or when I shall do it. My duty to do it is my duty to do this individual action and no alternative. And in the second place, there is nothing in this form of action corresponding to the question [of] why I choose this end and not another end or why I choose this rule and not another rule. Choosing my duty does not imply that it was an open question what duty I should choose, as choosing an end leaves it implie[d] that it was an open question what end I should choose, or as choosing a rule leaves it implie[d] that it was an open question what rule I should choose to obey. My consciousness of duty is a completely individualized consciousness; it springs from my consciousness of the situation in which, as a practical agent, I find myself or place myself.

But here arises a difficulty. The utilitarian analysis of an action divides it into two actions, means and end. It explains, or partially explains, why I will the means by referring to the fact that I will the end. The regularian analysis of an action divides it in a different way into two actions, willing the rule and willing to obey the rule. It explains, or partially explains, why I will to obey the rule by referring to the fact that I will the rule. Of the three answers to the question 'Why do I choose this?', the first and second are derived from analysis of what I do into two parts, and explain, or partially explain, one part by reference to the other. But the third answer involves no such analysis. It says 'I do this because it is my duty.' But what I do is identical with my duty; and therefore, if the reason for a thing must be a different thing, as I said in an earlier lecture, the proposition 'I do this because it is my duty' gives no reason for my doing it. It is a tautology, and the answer it gives to the question 'Why do I do this?' is an unreal answer.[2] It amounts only to saying 'I do this because I have got to do it,' and adding, 'I have got to do it because I have got to do it.'

At this point, any of you who have been attending critically to my argument might be tempted, if the manners of a lecture-audience did not restrain him, to intervene in some such terms as these. 'See what comes', you might say, 'of trying to go beyond the customary utilitarian and regularian ethical theories. Utilitarianism and regularianism offer you half a loaf; they offer you a partial explanation of moral action. Because this explanation is only partial, you have attempted to improve on it, and expound a conception of duty which shall provide a complete explanation. Now you find that the complete explanation, just because it is complete, is no explanation at all. You would have been wiser to remain content with explanations which, although only partial, did at least partly explain.'

---

[2] The argument to which he refers is included in chap. 4 of this book.

I do not propose to take this advice. I admit having based my account of duty on something that has turned out to be tautological; but I have already said that a tautology may have, if not a logical value, at any rate a rhetorical value. And I propose to consider my own tautology from a rhetorical point of view. In an earlier lecture I said that the tautology 'I love you because I love you,' expresses a consciousness of the discovery that love is not a thing for which reasons can be given. This does not imply that there is no such thing as love. On the contrary, it implies that the speaker knows by experience both that there is such a thing and also what kind of a thing it is. So I claim for my own tautology 'I choose to do my duty because I have got to do what I have got to do,' implies that the speaker knows by experience both that there is such a thing as duty and also what kind of a thing it is.

A person who can say 'doing your duty is doing what you have got to do because you have got to do it,' is, I grant, speaking in tautologies. But they are tautologies symptomatic of a person who knows by experience that there is a form of moral consciousness which does not consist in analysing one's action into means and end, or rule and obedience to rule, or in any other way whatever, but is aware of it in its unbroken or unanalysed individuality as his own response to a situation in which, as I said, he finds himself or places himself. The situation is present to his consciousness just as the action is present to his consciousness, in its unbroken or unanalysed individuality. It is not present to his consciousness as a situation of such-and-such a kind or as a situation presenting such-and-such features, but as this situation and no other. His response to it, similarly, is present to his consciousness not as a response of a certain kind relevant to any situation of a certain kind or to certain features in this situation, but as this response and no other, relevant to this situation and no other, a response which is an *individuum omnimode determinatum* to a situation which is an *individuum omnimode determinatum*.

The consciousness of duty is thus the agent's consciousness of his action as a unique individual action relevant to a unique individual situation. This does not mean that the consciousness of duty is (*A*) the consciousness of an individual situation, and (*B*) the consciousness of an individual action relevant or appropriate to it. It does not mean that the consciousness of duty is, first, the consciousness of being in this unique situation, and secondly the consciousness of having to do this unique action because you are in this unique situation. If anyone should say that this relation existed between the two consciousnesses, he would expose himself to the valid retort that the opposite relation existed: that a man is aware of his situation as a unique situation only because he is aware that he reacts uniquely to it. Further reflection would show that there is here no relation of ground and consequent. My consciousness of the situation as unique is the same as my consciousness of my action as unique.

The fact of duty, then, is the fact that an agent is sometimes aware of himself as doing an action which is an individual action, if you like to put it that way; or if you prefer to put it this way, the fact that an agent is sometimes aware of himself as acting in an individual situation. This is the fact to which I was calling attention by the tautology 'I have got to do this because I have got to do it.' I wish to suggest that in describing this fact as the fact of conscious individuality, the fact that an agent is aware of his action or his situation as unique, I have been equating it with something that is already familiar to you in another context.

There is a thing called history. History in the ordinary current sense of the word is the name of a kind of thought whereby we come to know the past. But, as any historian will tell you, the past which history enables us to know is a peculiar kind of past. First, it is a past composed of individual events. History is knowledge of the individual. Secondly, it is a past composed not merely of individual events but of individual actions done by human beings in individual situations. The historian is able to find out about those actions and those situations only because the agents who did the actions were aware of the actions and of the situations. And the means by which the historian finds out about these things is his own awareness of his own situation as an historian, that is, his awareness of himself as having before him documents, or testimonies, or evidences which tell him about the past; and this awareness of his situation is also awareness of his acting in that situation, namely his activity as interpreting this evidence; for nothing is evidence to an historian except what he can interpret as evidence, and everything is evidence which he can interpret as evidence.

Every situation which the historian studies is an individual situation; every action is an individual action. Suppose, for example, he is studying the French Revolution. What makes him an historian is the fact that he thinks of the French Revolution as something unique; something of which there is and can be only one in the whole of history, something that did not and could not happen twice: not an instance of things called revolutions which have often happened and will often happen again, but the French Revolution. Other people may generalize about revolutions, but not the historians. This is not his business. His business is solely with the individual situation or complex of situations in its context of equally individual situations; the individual action or complex of actions in its context of equally individual actions.

And every historian is conscious not only of this; he is also conscious that his own interpretation or reconstruction or view of the French Revolution is itself an historical event. It is a unique view of the French Revolution: a view which could be taken only now, when the situation as regards documents and co. is what it now is, and when these documents are interpreted as they can now and only now be interpreted. Other historians at other times have taken different views of the French Revolution: they had to take different views, because the available documents were different (that is, because the historian stood in a different situation) or because the historian's outlook on the evidence was different (that is, because the action of interpreting the evidence was a different action). His historical consciousness is a consciousness not only of his subject in the past, but of his situation and activity in the present.

Finally, the historical consciousness is consciousness of a necessity which cannot be stated analytically in terms of reasons and consequences. When the historian has formed his view of the French Revolution as a complex of situations or actions in the past, he sees that this complex as a whole was a complex in which every detail had to be what it was. Why had it to be what it was? Because the French Revolution as a whole was what it was. The agents whose actions made up that whole which is called the French Revolution did what they had to do. They did not act capriciously. In fact, they acted in a utilitarian manner, discovering means to their ends. In fact, they acted in a regularian manner, so acting as to obey the rules they recognized. But what ends they pursued, and what rules they recognized, are questions whose only answer is: 'They did pursue those ends. They did recognize those rules. The fact that they did so is the fact that they were the men who made the French Revolution.'

Similarly, when the historian recognizes the historical character of his own work, he recognizes that in doing this work he has yet to do what he has got to do. He is not concerned to ask why the documents to which he has access are the documents to which he has access. That, for him, is simply an historical fact. Like other historical facts, it is a fact whose history he [can] trace. He can find out *how* these documents and no others came to be accessible; but not *why*, unless 'why' merely means 'how'. He is not concerned to ask why his own historical outlook is what it is. That, for him, is simply an historical fact. He is the historian he is: he has got to be the historian he has got to be. Once more, he can ask how he came by his own historical equipment, that is, he can find out how the historical fact of there being this historical equipment existing here and now in his own person came into existence; but not why it came into existence, unless 'why' merely means 'how'.

The consciousness of duty is thus identical with the historical consciousness. Now the historical consciousness, as I have briefly described it, first came into existence in the second half of the eighteenth century, that is to say, in the time of Kant. That is a statement on which I shall not here enlarge. I propose to enlarge on it in my lectures of next term, called 'The Idea of History'; in which I mean to trace the origin and development of that idea from ancient times down to the present.[3] The idea of history is the idea of action as individual; and this idea has itself had a long history, in which the late eighteenth century was the critical period. Even today, that idea has not attained the full development for which we may hope; and this is the same as saying that the idea of duty has not yet attained the full development for which we may hope. There has been in the past a utilitarian idea of history, and there has been in the past a regularian idea of history. But modern historical thought has left these ideas behind, or rather, has partially left them behind and is working today at leaving them completely behind. The modern idea of history as the individual historian's individual reconstruction of individual situations and actions in the past is still rather a promise than a performance; an idea which even now is not quite clearly and quite unquestioningly grasped as an achievement of thought. Historians are still writing, and still influential, who

[3] These lectures were given during the Trinity term (April–June) of 1940. They were, for the most part, written in 1936 and slightly revised in 1940. The lectures were edited and published posthumously as parts I–V, *The Idea of History* (Oxford, Oxford University Press, 1946).

analyse the individuality of past actions either by the utilitarian analysis or by the regularian analysis. Until these obsolescent tendencies in historical thinking have become wholly obsolete, we shall not have a fully developed consciousness of duty; or, if you prefer to put it the other way round, until we have fully developed consciousness of individuality in our own actions, that is, of duty, we shall not have a full understanding of the actions of men in the past as individual actions.

Consciousness of an action as an individual action does not preclude the possibility of analysing it by the utilitarian or regularian method. On the contrary, it removes from these analyses the residue of irrationality, or caprice which by themselves these methods of analysis are powerless to remove. When an action is analysed as an example of utility, as we have seen, two relics of caprice remain. First, the choice of an end is capricious. Secondly, the choice of one among various possible means of realizing that end is capricious. Now, if the action as a whole is conceived in its individuality, the end is no longer regarded as capriciously chosen; it is regarded as the end which one chooses because one has got to choose it. Similarly, the means by which one realizes it is no longer regarded as one possible means capriciously chosen from many various possible means; it is regarded as the means which one chooses because one has got to choose it. How these individual ends and individual means come to be chosen by a given agent in a given situation is a legitimate question for the historian. Granted that Gladstone was the man he was, conscious of himself as standing in the situation in which he was aware of himself as standing, the historian is able to ask how he came, towards the end of his life, to pursue Irish Home Rule as an end, and to pursue it, though unsuccessfully, through the means of parliamentary action. And these questions are historically answerable. In the same way, the double element of caprice that is involved in regularian action, caprice in the recognition of a certain rule, and caprice in the choice of one among alternative possible ways of obeying that rule, is removable by historical inquiry into the question [of] how it came about that a certain agent on a certain occasion chose to obey that rule and not a different rule, and to obey it in that way rather than another. The historian may hope to show that he recognized that rule because he had to recognize that rule and obeyed it in that unique way because he had to obey it in that unique way.

From this point of view we can both accept and expand the Kantian saying that 'ought implies can'. It would be a misunderstanding of that statement if we took it to mean that, first, one is conscious of various possible alternative actions, and that secondly one is conscious of a duty to choose one from among these alternatives. The consciousness of duty precludes the consciousness of any possible alternatives. In being conscious that we ought to do a certain individual action we are conscious of our ability to do it, and also conscious of our inability to do anything else. In being conscious of our duty we are conscious that this and nothing else is what we can do.

PART TWO:
Civilization and its Enemies

Editor's Comments

The overall theme of Part Two is the idea of liberalism, which is equated with freedom and civilization, and ultimately derived from the teachings of the Christian religion. Collingwood believes that liberalism, and hence European civilization, is under attack from all sides; not only from the Left and Right in politics, but also from the more general intellectual trends manifest during the age, but having their origins in earlier times. Here we see Collingwood addressing the substantive practical problems of the age which presented themselves for theoretical clarification. 'The Present Need of a Philosophy' and 'The Rules of Life' are included here to indicate the role Collingwood envisaged for the theorist in relation to the pressing social and political problems undermining civilization. The theorist is neither a disinterested party, nor one who can provide solutions to the problems. His or her role is one of inspiring confidence in the possibility of achieving answers to the most perplexing questions by clearly articulating the principle that, whatever evils may be associated with human institutions, there is none that the will cannot overcome. When this is translated into concrete advice, we see, for example, in relation to the proliferation of new movements sharing the common element of a frantic search for a leader, that Collingwood recommends that the individual first attains the level of freedom of the will before offering himself or herself to a leader. The way of achieving this freedom is to know himself, to respect himself, and to ensure that in directing his activities the passions are always under control, not denied or suppressed, but acknowledged and appropriately channelled into his practical engagements. The simple message of 'The Rules of Life' is in fact a distillation of Collingwood's continuous emphasis upon the development of self-knowledge through the levels of consciousness, the fullest exposition of which appears in Part I of *The New Leviathan*.

The principles of liberalism are most adequately defined in 'Modern Politics', the introduction to de Ruggiero's *European Liberalism*, and 'Fascism and Nazism'. What is clear from these expositions of the principles of liberalism is that liberalism and parliamentary democracy are not necessarily co-extensive. Indeed, in the *Autobiography* Collingwood suggests that Britain's parliamentary democracy had become most illiberal. Liberalism is a spirit, an attitude of mind, or even a method, which may have implications for the institutional arrangements of a society, such as a free press, unimpeded access to all points of view of relevance to an issue, and the adequate provision of political education, but the actual institutions through which liberalism works may vary from place to place and from time to time. Fundamentally, liberalism is the determination to solve political problems dialectically by allowing and facilitating opposing views to be voiced with the intention of discerning an underlying unity of conception as a basis of action.

Although liberalism and freedom are correlative, they are not inevitable, nor when achieved are they self-preserving. The principles of liberalism are fragile, not suited to all political circumstances, and somewhat at the mercy of those who may subscribe to them but fail to apply them consistently. This inconsistency in application, Collingwood suggests in 'Modern Politics', was the reason why liberalism provoked severe attacks from both the Left and Right. In 'Fascism and Nazism,' the custodians of liberalism, that is the subscribers to liberal principles, are indicted, not for applying them inconsistently, but for not cherishing the principles passionately enough, that is, for having lost the emotional energy, the religious love of God, which sustains them. Whereas Fascism and Nazism have been able to harness the emotional power of pagan religions, liberalism has had its emotional energy filtered out by 'Illuminism' and its heirs who seek to purge all elements of superstition and magic from the Christian religion leaving only a logical formulation of the liberal principles derived from it. This discussion is a development of that concerning the insidious influence of positivistic metaphysics to be found in *An Essay on Metaphysics*, and the implications of the utilitarian ethic associated with it are described in the tantalizingly brief 'The Utilitarian Civilization'. Its emphasis upon the futile attempt to suppress the emotional elements in civilization by understanding everything in terms of utility reinforces three points which are developed elsewhere. First, that such a false suppression of the emotions leads to a false conception of oneself, and having this false conception one tries to live up to it (*Speculum Mentis*). Secondly, the expression of emotion is vital to intellectual development and sane living. To deny these emotions is to corrupt the consciousness and pervert the foundations upon which the intellect builds. Superstition and magic, even though our civilization denies and ridicules them, are means by which we express emotion (*Principles of Art*). And, thirdly, of three levels of practical reason, utility, right, and duty, this extract from the fairy-tales manuscript, written in 1936, shows that Collingwood believed that our civilization was predominantly permeated by the guiding principles of the first level.

'The Prussian Philosophy', written shortly after the First World War, and predating 'Modern Politics', supplements the explanation given in the latter why liberalism was being besieged by forces from the Left and Right. The Prussian philosophy, Collingwood contends, gave rise to the modern absolutist theory of the state, and to the idea of the dictatorship of the proletariat, both of which constitute a grave threat to civilization. The penultimate article attempts to give a 'scientific' account of politics in terms of three laws which are universal in their application, but not self-enforcing. In particular, the conditions are explored which give rise to the reverse action of the third law, and serve to explain the effect of such tyrannical phenomena as Hitler and Mussolini on the body politic. The final extract, which is the original conclusion to *The New Leviathan*, outlines the criticisms offered against civilization and highlights the need to stem the tide of the revolt by acting positively. Collingwood's avowed intention to articulate the bare minimum of what every person must know in order to respond to the revolt against civilization is itself an indirect justification for bringing to light, and making more generally available, some of the discussions which have until now been condemned to languish in obscurity in the unpublished manuscripts.

II

The Present Need of a Philosophy

Your invitation to continue the correspondence opened by Sir Herbert Samuel[1] in the last issue of the Journal is one which I cannot in honour refuse; and I am the less reluctant to accept it, because the President's letter has expressed so many of my own convictions that I can follow his lead where I should have hesitated to venture alone. That philosophy ought in some way to help our generation in its moral, social, and political troubles; that epistemology and the theory of value are not directly contributing to that end; and that in this respect some special significance attaches to the idea of evolution—all this I fully and gladly accept; and I will try to say, as briefly as I can, what it is that in my opinion philosophy can do.

But first, there is something which it cannot, and must not be tempted to do. It cannot descend like a *deus ex machina* upon the stage of practical life and, out of its superior insight into the nature of things, dictate the correct solution for this or that problem in morals, economic organization, or international politics. There is nothing in a philosopher's special work qualifying him to pilot a perplexed generation through those rocks and shoals. If a mariner finds himself at sea without navigator, chart, or compass, the Astronomer Royal himself, discovered among the passengers, could do little for him; he would be wiser to hail some coastwise fisherman. Even Plato did not think otherwise. He never proposed that professional philosophers should be dragged, blinking, from their studies and forcibly seated on thrones; only that expert knowledge of political life and its practical difficulties should be illuminated by philosophical reflection on its ultimate end.

If nowadays we should hesitate to go even as far as Plato, it is not

Reprinted from *Philosophy*, 9 (1935), 262–65, with the kind permission of Cambridge University Press.

[1] Sir Herbert Samuel, the President of the Institute of Philosophy, was invited by the editor of *Philosophy* to generate a discussion of contemporary issues in the journal. Samuel impressed upon philosophers, 'the need of an effort more effective than hitherto, to give direction in these difficult times to a troubled world'. 'The Present Need of a Philosophy', *Philosophy*, 9 (1935), 134.

because our opinion of philosophy is lower, but because our opinion of the plain man is higher. Christian theology holds that the faith of a simple peasant, without any tincture of theological learning, is sufficient for salvation; modern philosophy, of whatever school, follows its example in holding that non-philosophical thought in all its forms—moral and political, scientific, religious, or artistic—is able to do its work without asking philosophy's help and to justify itself without awaiting philosophy's verdict.

In this opinion there lurks an opposite danger. It may seem that philosophy's only task is to analyse knowledge we already possess, and theorize about activities we are already able to perform; that it is no more able to influence the processes which it describes than astronomy can influence the movements of the stars; that the only motive to pursue it is a pure disinterested curiosity, the only good to be gained from it, pure theoretical knowledge; and that Plato, Spinoza, and all others who have thought this knowledge somehow serviceable to our well-being were victims of a gigantic and inexplicable illusion.

The truth seems to me to lie somewhere between these two extremes. If the philosopher is no pilot, neither is he a mere spectator, watching the ship from his study window. He is one of the crew; but what, as such, is his function? In order to find an answer to this question, I suggest that we should look back three hundred years or more, to the infancy of modern science. At the beginning of the seventeenth century no one could foresee the triumphs which science was one day to achieve. It was not, therefore, a foreknowledge of these triumphs that encouraged innumerable men to persevere in almost incredibly detailed inquiries concerning the laws of nature, in a corporate effort shared by all parts of the civilized world and extending over many generations. The will to pursue those inquiries was not based on any conception of their future outcome, but it was based on something: it was based on the belief that nature is a single system of things, controlled throughout its extent by a single system of laws. In adopting this idea, civilized man was setting aside his immemorial belief in demonic agencies, magical influences, and the inscrutable caprices of individual things, and accepting a new view of the world, not received on faith, and not arrived at by scientific induction, but thought out and stated in a systematic form by the philosophers of the sixteenth century.

The notion of a uniformly law-abiding natural world is so familiar to ourselves that we are apt to forget how recent a thing it is in the history of thought, how hardly it was won by Renaissance thinkers—for example, with what difficulty sixteenth-century thought gave up Aristotle's doctrine that the law of gravitation holds good only in the sublunary sphere—and how dramatic was its verification by one scientific discovery after another. This philosophical conception of nature has played the part, in relation to scientific research, of a constant stimulus to effort, a reasoned refutation of defeatism, a promise that all scientific problems are in principle soluble.

There is a certain analogy between the state of things at the beginning of the seventeenth century, when the special problems of civilized life were concerned with man's control over nature, and the state of things in the modern world, whose special problems are concerned with human relations. Sir Herbert Samuel justly enumerates them: 'personal and social morality, economic organization, international relationship'. These problems, like the problems of natural science, can be solved only by detailed and patient investigation, exhaustive inquiry, skilful experiment. But this arduous and slow labour, if it is to be undertaken at all, must rest on two things: a conviction that the problems can be solved, and a determination that they shall be solved. Of these two, the first is, I think, capable of being provided, in a reasoned form, by philosophy. Apart from such a reasoned conviction, the will to solve them is so handicapped by doubts within and opposition without, that its chance of success dwindles to vanishing point. There is always a vast mass of opinion (and very respectable opinion) in favour of allowing established institutions to stand firm for fear of worse to follow; there is always a dead weight of inclination, however bad things may be, to enjoy what good we can snatch for the short time allowed us; but, more dangerous than either of these, there is the defeatist spirit which fears that what we are aiming at is no more than a Utopian dream. And this fear becomes paralysing when, not content with the status of a natural timidity or temporary loss of nerve, it calls in the help of philosophical ideas, and argues that the evils admittedly belonging to our moral, social, and political life are essential elements in all human life, or in all civilizations, so that the special problems of the modern world are inherently insoluble. The philosophical ideas underlying this argument are connected with certain aspects of the idea of progress; especially the false conception of progress as due to a cosmic force which can be trusted to advance human life automatically, without the active co-operation of human beings, and (the natural reaction from this) an equally false denial that progress is possible at all.

As the seventeenth century needed a reasoned conviction that nature is intelligible and the problems of science in principle soluble, so the twentieth needs a reasoned conviction that human progress is possible and that the problems of moral and political life are in principle soluble. In both cases the need is one which only philosophy can supply. What is needed today is a philosophical reconsideration of the whole idea of progress or development, and especially its two main forms, 'evolution' in the world of nature and 'history' in the world of human affairs. What would correspond to the Renaissance conception of nature as a single intelligible system would be a philosophy showing that the human will is of a piece with nature in being genuinely creative, a *vera causa*, though singular in being consciously creative; that social and political institutions are creations of the human will, conserved by the same power which created them, and essentially plastic to its hand; and that therefore whatever evils they contain are in principle remediable. In short, the help which philosophy might give to our 'dissatisfied, anxious, apprehensive generation' would lie in a reasoned statement of the principle that there can be no evils in any human institution which human will cannot cure.

This cannot be done in a day. But it has already been well begun. I will mention three writers whose work, taken as a whole, seems to me unmistakably converging upon a conception of man and his place in the universe which would justify that principle. There is Mr Alexander's *Space, Time, and Deity*; there is General Smuts's *Holism and Evolution*; and there is Mr Whitehead's series of books grouped round *Process and Reality*. These, with others hardly less important, seem to me the first-fruits of a new philosophical movement in which epistemological discussions and the old controversy between realism and idealism have fallen, as Sir Herbert desires that they should fall, into the background; in which the central place is taken, as Sir Herbert wishes it should be, by the idea of development; in which philosophy feels itself a collaborator with science, neither its enemy nor its slave, but having its own dignity and its own methods, while it respects those of science; and in which man is conceived neither as lifted clean out of nature nor yet as the plaything of natural forces, but as sharing, and sharing to an eminent degree, in the creative power which constitutes the inward essence of all things.

12

The Rules of Life

In lecturing to you, I have not been trying merely to supply you with materials for writing a successful paper on Moral Philosophy in the Schools. Whether these lectures will do that for you or not, I am not sure; but at least they were not chiefly meant to do that, but to direct your thoughts towards the fundamental problems that are concerned with the life of action. The life of action is a life which we all have to lead; there is no escaping it, whether by retiring from the world into a contemplative life or by living in the world under the command and leadership of others to whom we entrust our souls; and the problems that it presents are not academic problems, arising only when we begin to think about it but non-existent in the living of it. Had that been so, moral philosophy would have been no fit study for men whose interest lies in action itself; but as it is, we cannot act without thinking what we are doing and why we do it, and clearness of thought about these matters is a condition of acting wisely and firmly.

It is inevitable, therefore, that in these lectures I should have been trying so to describe my own experience of action as to make it useful to you in your active lives. Everything I have said is meant to have some application to practice and to ease the task, so far as that task can ever be eased by other people's words, of deciding how to live. I should like to spend these last few minutes on showing you a little of what I mean.

You have found yourselves growing up in a world whose chief singularity is that nothing in it can be trusted to stand firm. Your parents were brought up in a framework of political and social ideas and institutions within which they could live their own lives in certainty as to what was required of them and in confidence that if they respected the ideas the institutions would protect them, give them security of life, and, what is more important, peace of mind. This framework has collapsed. The systems of social, economic, and

Extract from 'Lectures on Moral Philosophy, 1933', Collingwood MS DEP 8, pp. 127–30.

political order which at the beginning of the century seemed fully capable of protecting the person, property, and thoughts of the individual man or woman have been subjected to strains which leave them damaged, I will not say beyond repair, but certainly beyond the line of efficiency. Today we stand—I say we, for we are all in the same boat—naked and defenceless against forces from which your parents in their youth were adequately sheltered by these now shattered institutions. It is hardly to be expected that they can ever entirely adjust themselves to so momentous a change; accustomed to the well-lighted spaces of their youth, they cannot but grope uneasily in the unaccustomed darkness. It is for you to make yourselves at home in that darkness and learn to find your way about in it.

At first, after the war, people behaved as people always behave after a great catastrophe. Misguidedly and in a spirit of blind hope they set themselves to rebuild what had in fact been finally destroyed. Then, finding that this was not so easy, they rushed to the other extreme and plundered the ruins in which they lived, and called this having a good time. Now, in the last few years, I see a new spirit coming into existence. It is a spirit of serious constructive work, a spirit anxious to devise new means of accomplishing what is essentially a new task. New and highly experimental organizations are growing up among the young, some of them political, whether Communist or Fascist, some religious, like the groups that have taken their name from this University; all these, however opposed to each other, share in this new spirit that I have called constructive and experimental.

It is not my business here to say anything of these movements in detail. But there is one common feature of them to which I must call your attention. Everywhere you will hear people saying that they want a leader. Demand creates supply, and leaders are forthcoming. But before you swear allegiance to any of them, I would ask you to remember this. If you find a leader worth following, you must offer him followers worth leading. You can only do this if you are so far independent of any leader as to live in a way that deserves the name of living before you find him. How are you to do this in the chaos and darkness of the world into which you have been born?

I have been trying to answer that question in my lectures, and I must now try to show you what my answer comes to.

The first and greatest rule of life, as it seems to me, is the old rule, know yourself. You must take pains, when you want something, to find out what it is that you really want. If you do not, you will find later that you have been pursuing things under the mistaken belief that you want them, and making it impossible to gain the things you really want. You must take pains, when you are doing something, to find out what it is that you are really doing. When you choose a course of action, you must take pains to find out what are the real reasons for your choice. This sounds simple, but there are a thousand impediments in the way: fear and shame of our real desires and real motives, fear that we shall have to forgo something we greatly desire when we find out that there is something else we desire more, every kind of hypocrisy and cowardice, which make self-knowledge the hardest thing to compass. But in so far as we do achieve it, we are lifted above dependence on things outside ourselves, whether ready-made institutions or ready-made leaders, and we are certain of happiness, so far as happiness is a thing that man can find.

The second rule is, respect yourself. Resist the temptation to belittle human nature by writing it down to the lowest level. Think that every desire, every impulse, every feeling which you find in yourself, and therefore in other people, has a right to be there, and demands not to be repressed or hidden away from sight, but to be fitted somehow into the map of life. There are lower and higher elements in our nature; but the lower elements are not purposeless, and cannot be killed without fatal damage to the whole, nor ignored without fatal ignorance of the whole. Respect not only your reason but your passions; not only your conscious mind but your unconscious mind; not only your mind but your body.

The third rule is, orientate yourself. In all activity there is process or development, movement from one element to another: see that the arrow which in your mind's eye is drawn on the diagram of your activities points in the right way. For example: passions like fear and anger represent activity breaking down and disintegrating into emotions: therefore point your arrow away from anger and fear, and never do anything because you are angry or afraid. Again, love begins as a mere animal appetite, but points from the first towards an effort to realize a perfect human nature, and in the last resort points to the love of God. Never, unless you wish to paralyse your will and frustrate your deepest desires, invert the order of things and treat love as a mere animal appetite to which these other things have become attached as a kind of fig-leaf to conceal the crudities of sex from prudish eyes.

With these three rules in mind, I would say to you, when you look for shelter behind institutions or leaders, don't look for help to things outside you. Look inside yourselves. Learn not to be frightened of what you may find there. Learn to look deeper and deeper, solving the perplexities you find at one level by penetrating to the one below. In a world where institutions have broken down and leaders have failed, this resource is still open to you; it is the resource men have always had in such times, and it has always been enough. If you can look deeply enough into yourselves, you will find there not only the means of living well in a disordered world, you will find, what you will never find elsewhere, the means of building a new world for your more fortunate children to inhabit.

13
Translator's Preface

The words *liberal* and *liberalism*, as used in this book, have a significance far wider than the platform or policy of any single party. They are used 'in their continental, not in their British, meaning. We borrowed them from abroad, and have used them to designate a party, or, rather, a particular section of a particular party. But "Liberalism" as used in its original home is a name for principles of constitutional liberty and representative government, which have long been the common property of all parties throughout the English-speaking portions of the world.' In these terms Lord Balfour explained his own use of the words in his introduction to the English translation of Treitschke's *Politics*,[1] and, with such authority, the present translator need not apologize for a severely literal rendering of his author's terms.

Liberalism, as Professor de Ruggiero understands it, begins with the recognition that men, do what we will, are free; that a man's acts are his own, spring from his own personality, and cannot be coerced. But this freedom is not possessed at birth; it is acquired by degrees as a man enters into the self-conscious possession of his personality through a life of discipline and moral progress. The aim of Liberalism is to assist the individual to discipline himself and achieve his own moral progress; renouncing the two opposite errors of forcing upon him a development for which he is inwardly unprepared, and leaving him alone, depriving him of that aid to progress which a political system, wisely designed and wisely administered, can give.

These principles lead in practice to a policy that may be called, in the sense above defined, Liberal; a policy which regards the State not as a vehicle of a superhuman wisdom or a superhuman power, but as the organ by which a people expresses whatever of

G. de Ruggiero, *The History of European Liberalism*, trans. R. G. Collingwood (Oxford, Oxford University Press, 1927). Reprinted with the kind permission of Oxford University Press.

[1] A. J. Balfour, introduction to von Treitschke, *Politics*, trans. B. Dugdale and T. de Bille (London, Constable, 1916).

political ability it can find, and breed, and train within itself. This is not democracy, or the rule of the mere majority; nor is it authoritarianism, or the irresponsible rule of those who, for whatever reason, hold power at a given moment. It is something between the two. Democratic in its respect for human liberty, it is authoritarian in the importance it attaches to the necessity for skilful and practised government. But it is no mere compromise; it has its own principles; and not only are these superior in practice to the abstractions of democracy and authoritarianism, but, when properly understood, they reveal themselves as more logical.

The development of this conception, in the political theory and practice of the last hundred years and more, is the subject of this book. It is a subject of the utmost interest today, when from various sides, in various countries, the political systems that take their stand upon freedom are being attacked by powerful and dangerous enemies. Is political freedom a chimera, or is it the one lodestar of sound policy? Is it destined to disappear, crushed between the opposing tyrannies of the majority and the minority, or has it the strength to outlive its opponents? To these questions our author has here given a reasoned answer, and one to which no thoughtful student of politics can be indifferent.

. . . .

14
Modern Politics

The plainest political fact of our times is the widespread collapse of what I shall call, using the word in its Continental sense, liberalism. The essence of this conception is, or was, the idea of a community as governing itself by fostering the free expression of all political opinions that take shape within it, and finding some means of reducing this multiplicity of opinions to a unity. How this is to be effected, is a secondary matter. One method is to assume that the political opinions held in the community fall into two main groups; to use this division as the basis for a distinction between two parties; to form a parliament whose members are elected by constituencies; and finally to form a government consisting of members of that party which has a majority in parliament. What makes such a system liberal is none of this machinery, but the principle that political decisions shall be made solely by persons who have heard, or at least had every opportunity to hear, both types of opinion about them. A constituency elects its member after hearing both candidates express their views on current issues; parliament, as the sovereign body, acts after a debate in which chosen spokesmen on both sides take part.

This is, of course, not the only possible system that is liberal in character. No necessary connection exists between liberal principles and territorial constituencies, or majority votes, or even parliamentary representation. The one essential of liberalism is the dialectical solution of all political problems: that is, their solution through the statement of opposing views and their free discussion until, beneath this opposition, their supporters have discovered some common ground on which to act.

The outward characteristic of all liberalism is the fact that it permits the free expression of opinion, no matter what the opinion may be, on all political questions. This attitude is not toleration; it is not the acquiescence in an evil whose suppression would be a greater evil; it is not a mere permission but an active fostering of

Extract from 'Man Goes Mad', Collingwood MS, DEP 24, pp. 16–28.

free speech as the basis of all healthy political life. Behind this outward characteristic lies an equally characteristic theory. It is implied that political activity and political education are inseparable, if not identical, that no politician is so expert that he should be allowed to govern without being exposed to the criticism of his fellow experts, none such a beginner or ignoramus that his opinion is valueless to the community. This again implies that politics is a normal form of human activity, one which can be shared very widely among mankind and ought, for the sake of the best human life, to be shared as widely as possible. Political activity is conceived as something that circulates freely through the community and draws vigour from this free circulation: good governments receiving additional strength from the enlightened and co-operative criticism of their subjects, so that, the more widely political experience and political thought are diffused, the better the government is likely to be. Thus the aim of a liberal system is not simply to solve the political problems of the community as they arise, but to act as a constant school of political experience for the entire body politic, training new leaders and training the rest to co-operate with them in their task.

This conception of political life has been gradually worked out in Europe and America during the last three centuries. It is not a monopoly of any one nation. Perhaps the chief contribution to it, both in theory and in practice, have been made in France, England, and the United States; but it can be properly understood only as a corporate effort on the part of our civilization as a whole; and it is certainly one of the greatest achievements of that civilization. But to call it an achievement is hardly accurate, for it is not and never has been a finished product; as a theory, it has never been authoritatively and finally stated, and as a form of political practice it has never been worked out in all its implications. It has been, not so much something done, as something in the doing: a course of action to which our civilization has committed itself, realizing that the developments to which it will lead are not wholly foreseen.

There are certain conditions under which alone liberalism can flourish. It is not the best method of government for a people at war or in a state of emergency: for then silence and discipline are demanded of the subject, bold and resolute command of the ruler. It is not the best method for a people internally rotten with crime and violence: there, a strong executive is the first thing needed:

force must be met by force. These restrictions, however, do not amount to criticisms of liberalism on its own ground. It professes to be a political method, that is, a method by which a community desiring a solution for its own political problems can find one. War is not part of politics, but the negation of politics, a parasitic growth upon political life. To say of a political system that it will not work under conditions of war is only to repeat that it is a political system. Similarly crime is a parasitic growth on the social structure. Society, it has been said, lives on the assumption that murder will not be committed, and for murder one may read crime in general. A community can only live a genuinely political life when crime is, not indeed abolished, but so far kept in check that for ordinary purposes it may be ignored.

Liberalism, then, requires for its success only one condition: namely that the civilization which adopts it shall as a whole and on the whole be resolved to live in peace and not at war, by honest labour and not by crime. It might seem, therefore, that liberalism is a mere utopianism, based on a blindly optimistic view of human nature. But this is not the case. A liberal government is still a government, and like every government must enforce law and suppress crime. Because it sets out to hear every political opinion, it is not committed to the dogma that every human being under its rule has such opinions.

At the present time, liberalism is undergoing attack from two sides at once, by two opposite parties and for two opposite reasons. First, there is the attack from the right. Here the complaint is that liberalism talks instead of acting. Instead of taking up definite problems and fitting them with definite solutions, it spends time collecting opinions about them. What it lacks is efficiency. The remedy is to suppress parties, parliament, and all the apparatus of a political dialectic, and to entrust the work of government to an expert, exempt from criticism and endowed with power to command, who shall invent his own solutions for all problems as they arise and impose them upon an obedient community.

The ground on which this doctrine rests betrays a genuine and absolute opposition to liberalism. The situation is represented as one of emergency. In emergencies, the method of liberalism is no longer valid. But what we are considering here is no temporary suspension of *habeas corpus* and the freedom of the press, it is a permanent declaration of a state of emergency. Naturally, this form of government is adopted most thoroughly in militaristic countries. Regarding war as the true end of the state, they have proclaimed martial law as a substitute for government, and reject liberalism not because it is politically unsound, but because it is political. The first French Republic fought for liberty and the rights of man; these gangster-governments suppress liberty and deny the rights of man, in order that they may fight.

But this attack on liberalism is gaining ground elsewhere than in countries avowedly militaristic. It is a poison that is permeating the whole of our civilization. A generation has elapsed since Lord Rosebery wrote upon his banner the ominous word 'efficiency'. During that generation there had been a ceaseless movement away from liberal principles in the direction of government by 'orders' originating in the offices of the Civil Service, and parliamentary government, with all that it stands for, has fallen increasingly into discredit. Today, members of all parties and shades of opinion may be heard expressing contempt for it is an outworn system, and demanding a more efficient and expeditious method of conducting public business. Even this country, the home of the parliamentary system, is visibly sliding towards authoritarian government.

This jettison of the liberal principles which our civilization so long and painfully acquired—a jettison conscious and violent in Germany and Italy, careless and almost absent-minded in our own country—is political madness. However much so-called liberal governments have failed in acting up to their principles, those principles themselves are the most precious possession that man has ever acquired in the field of politics. The conception of political life as permeating the whole community, of government as the political education of the people, is the only alternative to anarchy on the one hand and the rule of brute force on the other. The work of government is difficult enough in any case; it is only rendered possible if rulers can appeal, over the heads of criminals, to a body of public opinion sufficiently educated in politics to understand the wisdom of their acts. Authoritarian government, scorning the dialectic of political life in the name of efficiency, and imposing ready-made solutions on a passive people, is deliberately cutting off the branch on which it sits by de-educating its own subjects, creating round itself an atmosphere of ignorance and stupidity which ultimately will make its own work impossible, and make impossible even the rise of a better form of political life. Those enslaved populations, when the memories of a long liberal tradition have died out of their minds, will be no longer able to restore sound government. They will be able only to throw up one gangster government after another. So easy is it to destroy a fabric that has been centuries in the building.

The other attack on liberalism, from the left, complains in effect that liberalism, as it has actually existed, is not genuinely liberal at all, but hypocritically preaches what it does not practise. Behind a façade of liberal principles, the reality of political life has been a predatory system by which capitalists have plundered wage-earners. What is proudly described as the free contract of labour is a forced sale in which the vendor accepts a starvation wage; what is called the free expression of political opinion is a squabble between various sections of the exploiter class, which conspire to silence the exploited. Within the existing political system, therefore, the exploited class can hope for no redress. Its only remedy is to make open war on its oppressors, take political power into its own hands, establish a dictatorship of the proletariat as an emergency measure, and so bring about the existence of a classless society.

In one sense this programme is not an attack on liberalism but a vindication of it. The principles on which it is based are those of liberalism itself; and in so far as its analysis of historical fact is correct, it must carry conviction to anyone who is genuinely liberal in principle and not merely a partisan of the outward forms in which past liberalism has expressed itself. The correctness of this analysis has been demonstrated by the sequel. The attack on liberalism from the right has actually been the reaction of privileged classes to this challenge from the left. They have been confronted with a dilemma. The socialist has accused them of using liberal professions as a mask for exploitation. 'If you are liberal, as you claim to be,' he has argued, 'you will confess to exploiting us, and reform your social and economic system. If you cling to your privileges, you can no longer hold them in the name of liberalism: you must retain them, if at all, by brute force.' The attack upon liberalism from the right represents a realization by the privileged classes of the fact that this dilemma cannot be escaped; they have, to that extent, chosen the second alternative and dropped the mask of liberal principle to take up the weapons of class war.

But the socialist programme as I have stated it, though liberal in principle, is anti-liberal in method. Its method is that of the class-war and the dictatorship. Class-war is war, and the time is past when war could be waged as a predatory measure, in order to seize property or power held by another. That is the old conception of war, which, as we saw in the last chapter, no longer applies to the conditions of the modern world.[1] There, war means not the transference of property from the vanquished to the victor, but its destruction; not the seizure of political power, but the disintegration of the social structure on whose soundness the very existence of political power depends. In so far as socialists imagine the coming revolution as a war by which the proletariat shall take possession of what is now enjoyed by its oppressors, they are deluding themselves; the fabric of wealth and government would perish in the war. The only way of avoiding this result would be to ensure, by careful preparation in advance, that there shall be no war, that the revolution shall be accomplished without resistance. But as opinion in the privileged class hardens, reinforced by hope of aid from militaristic states outside, this possibility becomes more and more remote. The socialist revolution in Russia took the world by surprise: but the civil war that rages in Spain as I write, in which the aggressors have been not the socialists but their opponents, shows that socialism cannot expect another walk-over.

I have spoken of class-war as if it entered into the socialist programme as a project for a future undertaking; but no socialist will accept this description of it. I shall be reminded that the class-war is not something in the future, but something that exists here and now, and has existed for generations: a conflict, therefore, which there is no escaping, but one which must be fought out to a

---

[1] The conditions to which Collingwood refers are the traditional ideas that war may be conducted to protect non-combatants; that war is a means by which to bring about peace; and, the purpose of the armed forces of a nation is to protect the peaceful activities of the citizens. These ideas have become obsolete, Collingwood argues, and superseded by the insane obsession that the criterion of nationhood is the ability to wage war. No one stands to gain in modern war, and consequently greed as the motive for aggression has been replaced by fear. All combatants lose in war, but the defeated nation suffers the ultimate humiliation: the loss of its ability to make war, the very criterion of its identity. To avoid such humiliation every nation assumes the right to attack any other of which it is afraid before it becomes a victim of aggression itself. The modern nation has been transformed into a fighting machine, and the life of the nation identified with warfare; 'with the resulting disappearance of that peaceful life for whose protection the instruments of warfare were once thought to exist'. See 'Man Goes Mad', pp. 3–4.

finish. I have already shown the confusion upon which this doctrine rests: the confusion between the general idea of conflict or struggle, and the special form of struggle which is properly called war. Healthy political life, like all life, is conflict: but this conflict is political so long as it is dialectical, that is, carried on by parties which desire to find an agreement beyond or behind their differences. War is non-dialectical: a belligerent desires not to agree with his enemy but to silence him. A class-conflict within the limits of a liberal political system is dialectical: one carried on in the shape of class-war is non-dialectical. The ordinary socialist conception of class-war is equivocal, slipping unawares from one of these meanings to the other.

Dictatorship, again, no matter who wields it, means the imposition of ready-made political solutions on a passive people, and is the negation of liberal principles in political method. Orthodox socialism recognizes this, and regards the dictatorship of the proletariat as a brief emergency measure, to be dropped before it has brought about (as it must, if it continues) the political corruption of the whole community. But it is a dangerous matter to surrender principles for the sake of expediency. Only in so far as a people has no liberalism in its bones, can a dictatorship flourish in it for however short a time; and every day of that time means a further weakening of all liberal principle throughout the body politic. In Russia, accustomed as it was to Tsarist rule, the dictatorship of Lenin meant only the replacement of one autocracy by another; but in England, or France, or the United States a socialist dictatorship could be accepted only at the cost of an irremediable outrage to the deepest feelings of the people, a loss of that political self-respect which is the root and safeguard of all their healthy public life.

The fact is that socialism, in the Marxian form in which alone it is a vital force in the world today, carries along with it too much dead weight in the shape of relics from the age in which it was born. From the age of enlightened despotism it inherited the idea of a wise ruler imposing on the people his will for their good: a socialist dictator, placed upon the throne by a successful war, and inaugurating an age when production shall be organized and wealth distributed in the general interest. From the utopianism of the eighteenth century, which found perhaps its last expression in Kant's essay on *Universal History*, it inherited its dualism between a period of revolution and crisis, and a period when all conflict shall be at an end, the happy millennium of a classless society. From the romanticism of Hegel it inherited the idea of war—not war in general, but class-war—as a glorious consummation of political activity, the same idea which has maddened the brains of the militarists.

All these ideas are obsolete: they have been exploded once for all by that very liberalism against which they are now used as weapons. Enlightened despotism as a political ideal has yielded to the conception of a people as governing itself by a dialectic of political opinion. The dualism between a time of troubles and a millennium lying beyond it has yielded to the conception of conflict as a necessary element in all life and (as yet) not destroying its peace. The conception of war as at once glorious in itself and necessary to the achievement of human ends has yielded to the conception of war as distinct from conflict in general, as something anti-political and, so far as it is merely war, merely evil. In all these three ways socialism, in spite of its affiliation to Hegel's dialectic, shows itself radically un-dialectical, and it is liberalism that has proved the true heir of the dialectical method. To use a socialist phrase, they are undigested lumps of bourgeois ideology in the stomach of socialist thought.

If the abandonment of all attempt to live by liberal principles is madness, why has this madness come upon us? It is easy to blame the mad themselves: to argue that the left has abandoned liberalism because the working classes are resolved to seize property and power from their oppressors, and are too ill-instructed (their own leaders being blind leaders of the blind) to know that what they want to get must be destroyed in their attempts to get it, or else too much the slaves of envy and hatred to care, even though that should happen: to argue that the right has abandoned liberalism because the privileged classes are resolved to hold what they have, even though they and it must perish in defending it. Nothing is gained by blame: something perhaps, by trying to understand.

Liberalism, during the period of its growth and greatness, entirely transformed the inner political life of those countries where it took root. But it never applied itself seriously to the task of reforming their international relations. A statesman of the late nineteenth century, reading the narrative of the massacre of St Bartholomew or the treatment of the Dutch by Alva, would feel that he was reading of events in another world. It had become impossible for subjects of any civilized country to be so used by their government. But if he read of international events in the sixteenth and seventeenth centuries—the wars, the alliances, the treaties—he would recognize little or no change in the practice of his own age, as compared with that of which Hobbes wrote that 'Kings and persons of sovereign authority, because of their independency, are in continual jealousies, and in the state and posture of gladiators, having their weapons pointing, and their eyes fixed upon one another, that is, their forts, garrisons, and guns upon the frontiers of their kingdom, and continual spies upon their neighbours.'[2] What change there had been, was for the worse: weapons more destructive, wars more expensive, and national hatred (a thing hardly known in the seventeenth century) smouldering everywhere. The liberal state of the nineteenth century conceived itself as an individual among individuals, in that false sense of individuality which makes it synonymous with mutual exclusiveness, and denies that between one individual and another there may be organic relations such that the welfare of each is necessary to that of the other. The liberal government which 'trusted the people' hated and feared peoples other than its own. It was this unnatural union of internal liberalism with external illiberalism that led by way of international anarchy to the militarism of today.

If liberalism failed to affect international relations, it failed also in certain ways to affect the inner life of communities. A division was made, both in practice and in theoretical writings, between the public affairs of the community as a whole and the private affairs of its members. It was held that whereas a man's political opinions were of interest to the government, whose business it was to elicit them for its own guidance, his private actions, so long as he did nothing illegal, were his own concern. In practice, this meant that his life as a 'business' man was under no kind of control by the state, so that the economic life of the community was an anarchy as complete as international politics. This was tolerable in theory only because of the extraordinary doctrine, learnt from Adam Smith, that free pursuit of individual interest best subserved the interest of all; in practice it was soon found wholly intolerable, and the misery of the weaker, to which it gave rise, was the source of modern socialism.

The two attacks that are now being delivered on liberalism are

---

[2] Hobbes, *Leviathan*, p. 63.

thus the result of no misunderstanding or perversity; they are the result of a double failure, by the accredited representatives of liberalism in the nineteenth century, to apply their principles to regions where the application was urgently needed. The militarism and the revolutionary socialism which threaten to destroy civilization today are a just punishment for its crimes in the years of its greatness. They spring, not from weakness or falsity in the principles of liberalism itself, but from the failure of our grandfathers to put those principles consistently into practice. Where these attacks show symptoms of insanity is the fact that they are directed, not against the incomplete application of liberal principles, but against those principles themselves. For three hundred years, civilized man has been working out a liberal system of political method, applying it, bit by bit, to the various parts of his corporate life. Now, because the application has not proved exhaustive, because there are still some regions unreclaimed by this method, it seems that man has decided no longer to use it, but to throw it away as an ill-tempered child throws away a toy, to give up the attempt at living a political life, and to live in future the life of a gunman, the life of violence and lawlessness, the life which Hobbes, thinking he described the remote past only, and not the future, called solitary, poor, nasty, brutish, and short.[3]

[3] Hobbes, *Leviathan*, p. 62.

15
Fascism and Nazism

When travellers are overcome by cold, it is said, they lie down quite happily and die. They put up no fight for life. If they struggled, they would keep warm; but they no longer want to struggle. The cold in themselves takes away the will to fight against the cold around them.

This happens now and then to a civilization. The vital warmth at the heart of a civilization is what we call a religion. Religion is the passion which inspires a society to persevere in a certain way of life and to obey the rules which define it. Without a conviction that this way of life is a thing of absolute value, and that its rules must be obeyed at all costs, the rules become dead letters and the way of life a thing of the past. The civilization dies because the people to whom it belonged have lost faith in it. They have lost heart to keep it going. They no longer feel it as a thing of absolute value. They no longer have a religious sense of its rules as things which at all costs must be obeyed. Obedience degenerates into habit and by degrees the habit withers away.

This is how Greco-Roman civilization came to an end. It died because the people who inherited it lost heart to keep it going. It died because the religious passion that provided its driving force had ceased to exist. In the days of Marathon and of Cannae that passion was very much alive. In the days of the earlier Roman Empire it was moribund. In the days of the Lower Empire it was dead. Henceforth the Roman Empire was a corpse, galvanised from time to time into spasmodic activity but without life of its own.

Everybody knew this at the time. The great Emperor Julian not only knew it, he tried to cure it by reviving the 'pagan' religions. Others had known it ever since the first symptoms of decay had set in; from the fourth century before Christ in Greece, from the first century in Rome. When Christianity began to conquer the Greco-Roman world, people who knew that world to be moribund and

Reprinted from *Philosophy*, 15 (1940), 168–76, with the kind permission of Cambridge University Press.

knew that the cause was decay in pagan religion fancied that Christianity was killing Greco-Roman civilization and imagined that the destruction of Christianity would bring it to life again; not realizing that Christianity was conquering the world only because the pagan religions were already dead and had left a void which nothing could fill, except a new religion, as universal in its claims as the Roman Empire itself.

Pagan thinkers, knowing that Greco-Roman civilization was dying or dead because religious passion that inspired it had failed, could only choose between despairing of the future and trying (necessarily in vain) to redeem it by reviving that same passion. Christian thinkers, knowing the same thing, could despair of Greco-Roman civilization and at the same time hope for a new civilization whose driving force should be Christianity itself. That is the theme of Augustine's *De Civitate Dei*.

Their hope was justified. In the course of time the religious passion of Christianity built a new heaven and a new earth: the earth of modern politics and the heaven of modern physical science. This is not the place to describe their details, or even their principles; or to show how intimately those principles were connected with the doctrines of Christian theology, and how accurately in general outline they corresponded with the prophetic vision of the great Christian thinkers who wrote while Rome was falling.

Modern philosophy and science have admirably succeeded in a kind of distillation, by which the rational contents of the Christian faith have been separated out from the mass of superstitious ideas and magical rituals in which they were embodied. The 'spirit' of Christianity, issuing from this alembic of the intellect, has been bottled clear and cold and labelled 'principles of politics', 'principles of scientific method', and the like. The residue, consisting of superstitious ideas and magical rituals, has been thrown away.

Thus in the last two hundred years Christianity has suffered a curious double fate. Whatever in it is capable of logical formulation as a system of first principles has been analysed and codified and has come to function as the axioms upon which our sciences of nature and history, our practice in liberal economics and free or democratic politics—in short, all the things which make up our civilization, are built. But whatever is not capable of logical formulation, whatever is in the nature of religious emotion, passion, faith, has been progressively exterminated, partly by ridicule and partly by force, under the names of superstition and magic. The two processes have gone hand in hand. The same men who have been most eager to formulate the principles in which Christianity has trained them have been most active in suppressing the 'irrational' or emotional elements in Christianity itself.

It was in the eighteenth century, more especially in France and England, that the movement called 'Illuminism' grew up and spread like an epidemic among educated men. Its purpose was double: positively, to formulate and develop principles which the historian, when he studies them, finds to be derived from Christian sources; negatively, to wage war on every kind of superstition and magic, and in particular the superstitious and magical elements in Christianity itself, and on religion as such, identified with these superstitious and magical elements. Under the influence of Illuminism, the French revolutionaries inaugurated a cult of 'La Raison', and Auguste Comte a cult of 'L'humanité'; Shelley (miscalled a Romantic poet, if Romanticism be defined as a reaction against Illuminism) declaimed against priests, wrote *The Revolt of Islam*, and signed himself 'atheist' in visitors' books; and Gibbon, ignoring all the evidence, fabricated the fable that the Roman Empire fell not through its own religious decay but beneath the swords of the barbarians. It is only in our own generation that the long-known truth has been rediscovered. In spite of the Romantics (Blake wrote 'Mock on, mock on, Voltaire, Rousseau,'[*] but who cared for Blake?) Illuminism became the orthodoxy of the nineteenth century; so that an age that prided itself on its scientific attitude towards everything knowable developed in fact a scientific attitude towards everything except religion. Towards religion its proudly cherished attitude was in the main a purely emotional one: an attitude of unreasoning and unintelligent hostility.

The idea of freedom—free speech and free thought for everyone; free inquiry and free discussion in science, free inquiry and free discussion in politics—had been distilled from the body of Christian practice by a long chain of thinkers, of whom the Illuminists were no small part; and in that condition it was available, bottled and labelled, for nineteenth-century use. The first rule of the life and

[*] 'Mock on, mock on, Voltaire, Rousseau; | Mock on, mock on, 'tis all in vain! | You throw the sand against the wind, | And the wind blows it back again.' William Blake, MS. Notebook, *c.* 1804, *The Poems of William Blake*, ed. W. H. Stevenson (London, New York, Longman Norton, 1971), 481.

philosophy which on the Continent are called 'liberal' and in
England, to avoid confusion with the special programme and
principles of one single political party, 'democratic', is to apply that
idea, devotedly if not always wisely, to every detail of human
activity. But to a reader who studies any classical exposition of it,
like that of John Stuart Mill, the grounds of this devotion are far
from clear. They could not be stated either in terms of the
utilitarianism nor in terms of the intuitionism which were the two
fashionable moral theories of the nineteenth century. Not in terms
of utilitarianism, because liberty is or was an end in itself, an
absolute value; and the only values recognized by utilitarianism
were derivative values, the values of means. Not in terms of
intuitionism, because intuitionism gave no grounds for anything at
all: its only reason was the 'woman's reason', 'It is so because it is
so'.[†] Kant's statement of it as the categorical imperative in its
second form: 'treat human nature . . . as an end in itself, never
merely as a means',[‡] was clear and conclusive, but fell under
suspicion because it too evidently betrayed its origin in the pietistic
religion of Kant's childish training. He had not sufficiently purged
his moral principles of religious emotion.

The real ground for the 'liberal' or 'democratic' devotion to
freedom was religious love of a God who set an absolute value on
every individual human being. Free speech and free inquiry concerning political and scientific questions; free consent in issues
arising out of economic activity; free enjoyment of the produce won
by a man's own labour—the opposite of all tyranny and oppression,
exploitation and robbery—these were ideals based on the infinite
dignity or worth of the human individual; and this again was based
on the fact that God loved the human individual and Christ had
died for him. The doctrines concerning human nature on which
liberal or democratic practice was based were not empirically derived
from research into anthropological and psychological data; they
were a matter of faith; and these Christian doctrines were the source
from which they were derived.

Beliefs or habits long inculcated will survive for a time their
logical grounds. All over Europe, during the nineteenth century,
the grounds of the habits and beliefs called liberal or democratic

[†] Shakespeare, *Two Gentlemen of Verona*, I, ii, 24.
[‡] See I. Kant, *Groundwork of the Metaphysics of Morals*, trans. H. J. Paton (New
York, Harper Row, 1964), 101, for a similar statement.

were being destroyed by the anti-religious propaganda of Illuminism and its heirs. Yet the consequences that had been long ago foreseen by philosopher-statesmen like Berkeley did not appear to be happening; until, late in the century, it was complacently suggested, by men in whom the habits inculcated by Christianity had survived Christianity itself, that 'Christianity as a system of dogma' should be given up and 'Christianity as a system of morality' retained.

At the present time the liberal or democratic form of life has been openly, ostentatiously, and (it must be added) triumphantly attacked in at least three major European countries. In Italy, Germany, and Spain a new form of life has established itself, in defiance of the liberal or democratic tradition and in outspoken opposition to it. I say nothing of Russia because in the past Russia had never accepted the liberal or democratic ideals, and although Communism as preached by Marx and Lenin is hostile to those ideals, the USSR Constitution of 1936 contains a great deal that is in harmony with them. Moreover, I have no personal acquaintance with Russia, and I prefer to speak of what I know.

The point to which I wish to draw attention is this. Alike in Italy, in Germany, and in Spain, the vast majority of the population is sympathetic to the liberal-democratic ideals and hostile to the Fascist or Nazi minority that has seized power. And persons belonging to that majority know very well why power has been snatched from their hands. It is because their Fascist or Nazi opponents have somehow contrived to tap a source of energy which is closed to themselves. Fascist and Nazi activity exhibits a driving power, a psychological dynamism, which seems to be lacking from the activity of those who try to resist it. The anti-Fascists and anti-Nazis feel as if they were opposed, not to men, but to demons; and those of them who have analysed this feeling say with one accord that Fascism and Nazism have succeeded in evoking for their own service stores of emotional energy in their devotees which in their opponents are either latent or non-existent. Fascism and Nazism may be silly, but those who believe in them believe in them intensely, care enormously that they should win their fight, and win it because they so greatly care to win it. Liberalism or democracy may be wise, but the people who care for it do not care for it passionately enough to make it survive.

Fascism and Nazism owe their success to the emotional forces which they have at command. Alternative explanations are current, but none of them fits the facts.

Those who believe in the Marxist dogmas explain them as class-movements. They are not class-movements. They affect every class alike, and there is no one class whose special interests they advance or whose special point of view they express.

Those who believe in the omnipotence of big business explain them as a counter-revolution of big business against the 'Communist Menace'. It is true that by inventing an altogether imaginary 'Communist Menace' Fascism and Nazism have at critical moments rallied big business to their support; but the movement which has at times employed big business men as its mercenaries, and at times cast them off again, is one thing, and big business itself is another; no one can identify them without suffering from judicial blindness.

Those who believe in the omnipotence of propaganda explain them as due to the power of advertisement. For Germany, this explanation has been refuted once for all by Mr Peter Drucker,[1] who has shown that Nazism came into power when all the instruments of propaganda were in the hands of its opponents, that its own propagandist utterances have always shown a complete disregard of consistency, and that its most fervent supporters never believed them. The same is true, to my own knowledge, of Fascism. For Fascists and Nazis alike, propaganda is a useful weapon, but not an indispensable one. It is folly to identify a weapon with the hand that wields it.

Fascism and Nazism, then, are successful because they have the power of arousing emotion in their support. They can annihilate even the most widespread liberal-democratic opposition in their own countries because those who believe in them 'think with their blood', as the Nazis say: care intensely about their beliefs, and can therefore overwhelm the liberalism or democracy which, because its votaries have for two hundred years progressively purged it of emotional elements, has become as purely as possible what the French call *cérébral*, an affair of unemotional thinking. A man who thinks with his blood, even if what he thinks is silly, will always get the better of a man who thinks merely with his brains, even if what he thinks is wise.

---

[1] *The End of Economic Man* (London, 1939), especially page 14, bottom, and the pages following. I owe much to this book, but I do not accept everything it contains.

What liberals or democrats think, there can be no doubt of it, is wise. To think in that way is to take a long view. It means building up a social order whose scientific organization will always be capable of understanding as much of Nature as man need understand in order to advance his own welfare as one natural being among others; whose political organization will always be capable of resolving as much of the inevitable tensions and antagonisms between the various particular societies (nations, classes, and so forth) into which human society must always divide itself, as must be resolved in order to keep human society a going concern instead of a suicide club. It means renouncing all Utopias and all millenniums: all hope of omnipotence or omniscience, whether in relation to Nature or in relation to man. It means believing that human beings will never have solved all their problems, overcome all their difficulties, or settled all their quarrels; but that they need never lack the wits, the power, or the good will to solve the problems which at any given moment they most urgently need to solve.

What Fascists and Nazis believe, there can be no doubt of it, is silly. It means conjuring up a social order based on the superstitious adoration of individual leaders who, being human, are neither infallible nor immortal. It means not only neglecting inquiry into natural law, but stamping out the spirit of free thought and the practice of free expression upon which that inquiry depends. It means not only failing to resolve the inevitable tensions between one human society and another, but exacerbating them as a matter of principle, in order to intensify the emotional solidarity which they generate in the particular societies concerned. So, with regard to Nature, it means no longer cutting your coat according to your cloth, and providing your own cloth for the coat you need, but plundering the cloth from someone else's stores. With regard to man, it means living in a constant atmosphere of terrorism and terror, quarrels and fear of quarrels: in short, neglect of Hobbes' First law of Nature and aspiration towards a state in which there is 'no place for Industry, . . . and consequently no Culture of the Earth; no Navigation . . . ; no commodious Building, no knowledge of the Face of the Earth; . . . No Arts, no Letters, no Society; and which is worst of all, continuous feare, and danger of violent death; And the life of man, solitary, poore, nasty, brutish, and short.'

If, in spite of this, Fascism and Nazism have conquered countries which together contain one-quarter of the entire population of Europe, or one-third of the population of Europe if the USSR is left out, the reason can only be that in their original homes, Italy and Germany, special causes have led to the abandonment of liberal or democratic principles. These principles, as we have seen, are derived from Christianity; and the emotional force, the 'drive' or 'punch', that once made them victorious was due to Christianity itself as a system of religious practice rich in the superstitious or magical elements which, in Christianity as in every other religion, generate the emotion that gives men the power to obey a set of rules and thus bring into existence a specific form of life. We have seen that Illuminism and other anti-religious movements have long ago exhausted liberalism or democracy of this emotional force and reduced it to a mere matter of habit. A mere habit has no 'punch'. It holds together with a success directly proportional to its own tenacity, which depends chiefly on the completeness with which it has been established, and inversely proportional to the strength of the destructive forces operating upon it.

All over the world liberal or democratic principles, having lost their 'punch' and having become mere matters of habit, have lost their initiative and have been thrown on the defensive. All over the world destructive forces are operating upon them. But in Italy and Germany they have only been established in a somewhat experimental spirit at a somewhat recent date, among peoples whose history had for many centuries been conducted without them. In England and France they are hard-won achievements, the fruit of continuous political experience and political education extending over many hundreds of years. In the United States they presided at the birth of a nation, and are part and parcel of its heritage. Such habits are hard to break; yet in all three countries they have shown signs of weakening, and it is a serious question how long their defence will hold out against the attack of forces which I will not here attempt to analyse.

The destructive forces operating against liberalism or democracy in Italy and Germany are often identified with the after-effects of the war of 1914–18. Germany had to undergo the humiliation imposed on her by the Treaty of Versailles; Italy smarted under the disappointment of hopes which that same treaty did not fulfil. I do not deny this; but I suggest that it does not suffice to explain the facts. The spark in a fuse is not the sole cause of an explosion.

Liberal or democratic principles, I have argued, are a function of Christianity. Consider the religious situation of Italy and Germany. The Christianity of Italy is very ancient and very deeply rooted; but it is a highly syncretistic Christianity. When first it became the official faith of the country it was a Caesaro-Papism which combined the principles of Christianity with the principles of a three-centuries-old Emperor-worship, the worship of a 'Leader' invested with divine attributes and venerated in a magical ritual. And even Augustus did not invent this worship of a human leader; it was deeply rooted long before this time in the practice of Mediterranean peoples. Fascism harks back consciously to the Roman Empire, and the various 'Fascist Decalogues' that have found their way into print, however they may differ in other points, agree in reasserting the formula of Emperor-worship, 'The Duce is always right'.[5] And the vitality of pre-Christian religion among the Italian people is abundantly proved by the numbers of pagan cults which have always been, and still are, fervently practised by them, not (like the pagan survivals of the seventeenth and eighteenth centuries in this country) under the Church's ban, but countenanced and consecrated by the Church itself and presided over by the Church's own priests.

Fascism as a system of political principles is a function, not of Italian Christianity, but of the pre-Christian paganism which has survived under the toleration and protection of Italian Christianity; and this is what has made possible an understanding between Fascism and the Vatican. The 'punch' of Fascism is derived from the pagan emotions which the Vatican has always tolerated. It surrenders nothing, therefore, in tolerating Fascism.

Protestantism from the first has persecuted these pagan survivals and has refused to incorporate them in its own conception of Christianity. But even in England, which has been officially a Christian country for seventeen centuries, they still exist, as everyone knows who is at all intimately acquainted with the life of an English village. Germany was the last European country to adopt Christianity; and in Germany the pagan survivals have always been extremely vigorous. Officially, instead of being tolerated as in Italy, they have been disowned; but unofficially they have been very much alive. It is no wonder, then, that they should now have come to the

[5] Articles 8 (of the 1934 Decalogue) and 10 (of that of 1938), state; 'Mussolini is always right.' See M. Oakeshott, *The Social and Political Doctrines of Contemporary Europe* (New York, Macmillan, 1947: first published 1939), 180–1.

surface; that Nazism should actively promote them; and that leaders of German Christianity should become opponents of Nazism and should find themselves in concentration camps.

Thus in Germany, as in Italy, though with very different consequences for the representatives of Christianity, the new political movement contains ideas drawn from the survivals of an unextinguished pre-Christian religion, and derives its 'punch' from the emotional appeal of that religion. The same pre-Christian religions, or similar ones, survive in all European countries, and even, unless I am misinformed, in the United States of America. In all these countries, therefore, the explosives exist which in Germany and Italy have shattered the inert façade of the principles derived from Christianity. Elsewhere these explosives may not exist in such large quantities; and the spark which has exploded them may, possibly, not be forthcoming. But it would be folly to forget their existence; and it would be criminal in those who know the facts to conceal their knowledge.

The time has long gone by when anyone who claims the title of philosopher can think of religion as a superfluity for the educated and an 'opiate for the masses'. It is the only known explosive in the economy of that delicate internal-combustion engine, the human mind. Peoples rich in religious energy can overcome all obstacles and attain any height in the scale of civilization. Peoples that have reached the top of a hill by the wise use of religious energy may then decide to do without it; they can still move, but they can only move downhill, and when they come to the bottom of the hill they stop.[2]

[2] This article was written immediately after reading Mr Joad's *Appeal to Philosophers*. Mr Joad asks for a truce of philosophical sects, a pause in logic-chopping, and an attempt to make fresh contact with the classical tradition in philosophy. The philosophers of the classical tradition, as I understand it, were men who used their trained faculties of thinking in order to think about facts, and primarily about facts of practical importance in relation to the lives of their fellow-men. No facts, in my opinion, are of greater practical importance at the present time than Fascism and Nazism. Our own country is fighting Nazis Germany. My friend, Mr Hooper, asks me for an article on 'some urgent important theme'. The most urgent theme I can think of is the necessity for taking Fascism and Nazism seriously; to stop flattering ourselves with the belief that they are baseless follies indulged by unaccountable foreigners, or the alternative belief that they are good examples which we should be wise to follow. What our soldiers, and sailors, and airmen have to fight, our philosophers have to understand. If I have understood them badly, I hope someone who understands them better will correct me. [C. E. M. Joad's article on 'Appeal to Philosophers' appeared in *Philosophy*, 15 (1940), 400–16.]

16
The Utilitarian Civilization

Ghost stories are common enough in our civilization to convince us that the psychological causes which produce them are still operative there. But among ourselves they are the exception rather than the rule. This can only be due to some special feature of our civilization. What can this feature be?

If magic in general is an expression of emotion, and grows up naturally and inevitably from its emotional roots, its disappearance in a given society would seem to indicate that, in that society, emotional vulnerability has been (partially at least) overcome by the deliberate cultivation of a thick-skinned or insensitive attitude towards emotion itself. This is actually the way in which our civilization has suppressed magic. After a long and hideous experiment in suppressing it by force, by burning witches, we came to see that burning witches meant believing in them, and that their victims' belief in them, what I have called emotional vulnerability, was the source of their power. So we changed our own attitude towards them: replaced persecution by ridicule, and gradually developed a whole system of education and social life based on the principle that magic was not a crime but a folly, whose success depended on a like folly in its victims.

This hard-headed, or thick-skinned, or rationalistic attitude towards life, which our civilization invented in the seventeenth century, worked out in the eighteenth, and applied to all aspects of human affairs in the nineteenth, is the dominant factor in modern civilization. The best single-word name for it is utilitarianism. Our civilization prides itself on being sensible, rational, businesslike; and all these are names for the same characteristic, namely the habit of justifying every act, every custom, every institution by showing its utility. The doctrine that utility is the only kind of value that a thing can have is called utilitarianism; and it is obvious to anyone who reflects on the general character of our civilization that it is, characteristically, a utilitarian civilization. We justify the state as a

Extract from [Fairy Tales], 'IV Magic', Collingwood MS, DEP 21, pp. 14–18.

means to the protection of life and property against crime; the armed forces of the state, as a means to protect these against foreign aggression; the church, as a means to inculcating sound morality; mechanized industry as a means of increasing wealth or saving labour; sport as a means of getting exercise and preserving health; clothes as a protection against the weather and a safeguard against the passions that are aroused by nudity; cleanliness as a protection against germs; and so forth.

This utilitarianism is more than a principle; it is an obsession. Whatever cannot be justified in this way our civilization tends on the whole to suppress. In general, it discountenances emotion and the expression of emotion; in particular, it distrusts art and religion as things not altogether respectable. To live within the scheme of modern European-American civilization involves doing a certain violence to one's emotional nature, treating emotion as a thing that must be repressed, a hostile force within us whose outbreaks are feared as destructive of civilized life. We have already had occasion to observe that our horror of savages is really a horror of something within ourselves which 'the savage' (that is, any civilization other than our own) symbolizes. We are now finding reason to think that this thing is emotion: for magic, which sums up all that we dislike in savage life, is beginning to reveal itself as the systematic and organized expression of emotion.

Accustomed as we are to explaining our own institutions on the utilitarian principle, we naively extend it without misgiving to those of other people. Opening *The Golden Bough* at random, in chapter VIII of *Adonis, Attis, Osiris*,[1] I read: 'When the earth quakes in some parts of Celebes, it is said that [all] the inhabitants of a village will rush out of their houses and grub up the grass by handfuls in order to attract the attention of the earth-spirit, who, feeling his hair thus torn up by the roots, will be painfully conscious that there are still people above ground.' Here a custom whose nature is obviously magical is explained in utilitarian terms. Such an explanation is called by psychologists a rationalization. It does not affect my point whether the rationalization was invented by the anthropologist who recorded the custom, or by the villagers themselves in answer to leading questions demanding an explanation of that kind, or even

---

[1] J. G. Frazer, *The Golden Bough*, IV, *Adonis, Attis, Osiris* (London, Macmillan, 1941: 3rd ed.), i. 200.

(as in a case previously mentioned) by Sir James Frazer himself; my point is that our anthropologists, looking at savage life from the point of view of the utilitarian obsession, require for this custom an explanation of this kind, and, having got it, are satisfied.

We, with our superior scientific equipment, know that pulling up the grass cannot stop the earthquake. These poor savages, who do it because they think it can, are thus acting under a delusion. Owing to a scientific error, they are doing an act as means to a certain end, when in fact it cannot further that end. Generalize this example, and you get what has become the orthodox definition of magic as pseudo-science. This definition of magic is the result of examining customs, whose true basis is emotional, through the spectacles of the utilitarian obsession.

But if this definition is wrong; if magical practices are not utilitarian activities based on scientific theories whether true or false, but spontaneous expressions of emotion whose utility, so far as they have any utility, lies in the fact that they resolve emotional conflicts in the agent and so readjust him to the practical life for which these conflicts render him unfitted; then a new problem arises about our own civilization. We pride ourselves on always acting from utilitarian motives or scientific theories; but that very pride should warn us that this belief about ourselves may perhaps be unjustified. We may be conceiving our own civilization not as it actually is, but as, with our utilitarian obsession, we should like it to be. We think that our rationalism has done away with magic because that is what we want to think; but is it true?

The question has a double importance. For the conduct of our own lives and the right ordering of our own societies, it is important that we should not live under illusions about the nature of our own civilization; but that is not the point with which we are here concerned. We are concerned to understand the mind of the 'savage', with its furniture of magical ideas; and we have already seen that unless we can sympathize with these ideas, by recognizing their kinship with certain elements in our own experience, we cannot hope to understand them.[2] If we are to understand the

[2] Collingwood argues, in earlier chapters of the MS from which this extract is taken, that the Philological, Functional and Psychological approaches to anthropology are all at root positivistic in that they view the savage as an object whose practices are considered, on the positivistic criteria applied to them, irrational. Only the historical method, which overcomes the mind–object dichotomy, instead of denying the savage,

'savage' mind, we must dispel our rationalistic conception not only of savage culture but of our own, and find among ourselves practices which fall under the conception of magic as we have now defined it.

acknowledges the savage within us, and at once understands the savage and self. See [Fairy Tales] B 'Three Methods of Approach: Philological, Functional, Psychological' and [Fairy Tales] C 'III. The Historical Method', DEP 21.

17
The Prussian Philosophy

. . . There is a good deal of talk going about to the effect that conquest, or imperialism as it is called, is fundamentally evil. It is not. Do not be afraid of the word imperialism. The right imperialism—the rule of the more civilized over the less civilized—is a necessary element in the education of mankind. Imagine what Europe would be without the discipline of Roman rule and the legacy of Roman law; and you can then imagine what the Africa and Asia of the future may yet owe to the imperialistic rule of European nations.

But imperialism as between one civilized state and another is a very different thing. True, the civilization of one nation differs from that of another, and each may pardonably believe its own to be the better. And probably it *is* the better, for it. But you cannot infer that it is therefore the best for another nation. We have seen, to take one instance only, Germany trying to impose her civilization upon the French—French in habits and mind, even if not in speech—of Alsace-Lorraine; and the result has been disastrous and terrible failure. And this must always happen when the relation between governing and governed races is transferred from its proper sphere to an improper sphere: from the relation between civilized men and barbarians to the relation between two civilized peoples.

Now I think it is obvious that this false and evil imperialism was the immediate cause of the war; and it is equally obvious that the most conspicuous examples of it have been the empires of Germany and Austria—Germany, where a great race tyrannized absolutely and ruthlessly, over fragments of smaller races which it had conquered—French, Poles, Swedes; and Austria, where Germans and

Extract from 'The Spiritual Basis of Reconstruction', Address to the Belgian Students' Conference at Fladbury, 10 May 1919. Collingwood MS, DEP 24, pp. 8–16. The address begins with an historical account of the role of Belgium in European affairs. Collingwood was well-qualified to give such an analysis to Belgian students because his work in the Intelligence Division of the Admiralty (1915–19) entailed producing intelligence manuals of the region; the most important were those on Belgium and Alsace-Lorraine, which included atlases. He also wrote the article on Luxemburg in the *Encyclopaedia Brittanica*, 12th ed. (1920), 811–12. See 'List of Work Done', Collingwood MS, DEP 22, pp. 16, 23.

Magyars combined to exert an equal tyranny over Czechs, Ruma-
nians, Jugoslavs, and Italians.

But our analysis must go deeper than this. Why were these
tyrannies exerted; what was the motive which inspired them? I
think the motive is to be sought in the characteristic theory of the
state which we have learnt to call Prussianism. This theory starts
from the undoubted fact that the individual man is by himself
powerless for good or evil: that as he owes his literal, physical life to
a social fact—the union of his father and mother—so he owes his
economic, political, and spiritual life to the society into which he is
born, the influences which shape his mind and the goods which he
can only enjoy by the co-operation of other men. From this fact the
Prussian philosophy argues to the conclusion that all originative and
creative power is vested in the state, and that the state is, therefore,
so to speak, God: a being answerable to no one but itself, and
possessing absolute right and absolute power over all its own
members, while owing no duties, no obligations, no responsibilities
towards any other state.

The falsity of this conclusion is evident from the fact that other
states do exist; and that, therefore, there exists a plurality of
absolutes—a conclusion which contradicts the first axiom of any
absolutist philosophy, which is that there can be only one absolute.
Consequently the Prussian form of absolutism stands self-
condemned.

Pardon me if I remind you of a few historical facts in connection
with the development of this strange philosophy. It found its first
expression with Hegel, though Hegel was too great a thinker to
believe in it entirely, and in his system it appears rather as an
irrational excrescence than as an integral part. And in the direct line
of descent it split into two branches—the monarchist form of
absolutism, and the socialist form: the one finding its absolute in a
personal ruler, the other (as formulated by Hegel's disciple Marx)
in the equally absolute dictatorship of the proletariat. And having
begun by denying the duties of a state towards all other states, it led
necessarily to the conception of the state as conquering and imperialistic. Towards other states the state is merely force, and the more
perfectly it is a state the more perfectly it shows itself as a force.
There is nothing one state can do to another except conquer it. The
will of the state can only be defined as a will to power. This is the

very formula of Nietzsche, merely following up the inexorable logic of the Prussian philosophy.

But even before Nietzsche made plain the crude message of that philosophy, the thought of Germany showed symptoms of its disease. Whenever a school of philosophy is sick of a mortal illness, this symptom never fails to appear. I mean pessimism. Pessimism is the reaction of the human intellect to a hopelessly false philosophy. Struggling to understand the world in terms of the false philosophy which it has learnt to believe, and finding it impossible, the mind is driven to despair: and pessimism is another word for despair. In the persons of Schopenhauer and Edouard von Hartmann, the German mind showed, to anyone who could understand, that it was suffering from a deadly disease; and Schopenhauer himself was conscious that this disease was the philosophy of Hegel.

It is this spiritual disease that has caused the war. The absolutist theory of the state, which drives every state that holds it into a career of aggression, and conquest, and tyranny—a career like that of a mad dog, only to be quieted in death—this theory was the root of the war from which we are now emerging, and it is only the eradication of this theory that can give us peace.

And do not let us be deceived by outward semblances. Marx believed in the theory as fully as Nietzsche, and Lenin believes it as fully as Bernhardi. The orthodox Marxian socialist stands in the direct line of descent from the founder of Prussianism: and his dictatorship of the proletariat, his class-war, are philosophically and morally on a level with the dictatorship of the conquering race and the holy war of the strong nation against the weak. There is no profit whatever in casting out the one if we are going to fall victims to the other. I say nothing against socialism in general: I do not know that collective ownership of the means of production may not be as good as private ownership, or better. But a socialism which is Prussianist in its glorification of aggression, of war, and of tyranny is part of the spiritual disease which I am trying to describe to you.

Such is the disease: how are we to cure it?

For this, I must ask you to follow yet another step of analysis. We must see deeper into the cause before we can suggest a remedy.

The fault of the Prussian philosophy is its conception of the state as mere power. Power unchecked by anything like international law or international morality: its conception of the state as responsible to nothing higher than itself, as having no duties, no obligations, no responsibilties to anyone except itself. In framing this conception, it was attempting to conceive the state in the likeness of God: it succeeded in creating a state that was rather an incarnation of the devil. But can we say why, exactly, this conception was so false and so disastrous?

I think the reason is this. In separating the positive idea of power from the negative idea of limitation of power the Prussian philosophy made a fatal error. Where power exists, it must always be limited: if not by the will of others, then by its own will. My power of action is limited, as long as I am a man among men, by the wills of other men; but even if I were free from all external compulsion, I should find it necessary to set limits to my own power from within, by fixing upon objects to work for and means to adopt in the work. In the freest and most obviously unfettered activity—the activity, say, of the creative artist—the mind is not lawless: it sets itself problems to solve and determines that they shall be solved in this or that way and no other. And where man had tried to conceive an absolutely omnipotent will, as Christian theologians have done, he has always believed that even God creates for himself a moral law which he will not break. Power without law cannot exist; it is only within the limits of law that power can operate at all. A power therefore that is above law is a contradiction in terms. If the state is infinitely powerful, it must necessarily be infinitely sensitive to obligations and responsibilities.

I think therefore that a clear and frank recognition of international law or international morality—call it which you will, for law is the letter and morality the spirit—is one of the remedies against Prussianism. And thus the instinct of the world is clearly right in demanding and creating a League of Nations as a safeguard against a revival of the Prussian spirit. But a League of Nations is not enough. An institution may be captured by the very spirit which it is designed to suppress: and a League of Nations might possibly become the instrument of the domination of race over race or class over class. The ideal of the League of Nations must be interpreted according to the spirit. It must be understood to mean the abolition not only of race tyranny but of class tyranny: and the prevention not only of national war but of class war. And this it can only do in one way: namely through the spirit of mutual co-operation and trust. Co-operation first; and trust afterwards: for so long as we do not actively help our neighbours we cannot demand that they should trust us. We must realize—each one of us individually—that there is no evil in our country or in the world that is not in some degree our personal concern; that it is our positive and urgent duty to help in the realization of every good not only for ourselves, our class, and our race, but for everyone.

It is not easy to co-operate with our neighbours. We may sink our differences in time of war, when a common and visible danger threatens all alike: but at other times we are chiefly conscious of the divergence and conflict of interests and desires, and the lessons of war are soon forgotten.

But this war is not over. The German Empire is defeated, but the Prussian philosophy is not crushed. It stares us in the face, menacing and powerful. It has been defeated in one form to reappear in another and a more dangerous shape. We know that if Germany had conquered Europe our own civilization, our own liberties, would have been destroyed. But the ideal of class-war may yet conquer Europe, and the result will be the same. Remember this: once class war begins, it will not matter who wins. The result in each case will be disaster and death, the destruction of civilization. The dictatorship of the proletariat will destroy many of the things which we, as students and followers of art and science, value most highly. The dictatorship of the capitalist will destroy things no less precious.

There is only one way to avoid this calamity, this triumph of false imperialism. We must remember that we civilized nations have still to exercise our function of true imperialism: to bring light to the dark places of the earth. If we can do that, our civilization will be justified in the eyes of history as the civilization of Rome has been justified. But we can only do it if we conquer the false imperialism of enslaving our fellows.

This, then, is my message to you. Let me put it as shortly and clearly as I can. We have a heritage of gifts—political, legal, scientific, artistic—in virtue of which we call ourselves civilized. Our task is to civilize the world. That is the task of Belgium: the task of each one of you. But meantime, we have the choice of either increasing the value of these gifts or of destroying them. We can increase them if we help every race, every language, every class among ourselves to develop its own gifts to the uttermost. This is liberty and co-operation: that is the immediate aim of all that is noble in our political systems. But we can very easily destroy these gifts. They may be by our fault so completely destroyed that Europe shall leave behind her nothing but the legendary glories of a vanished Atlantis, and the dust of buildings like the palaces of Babylon. And this is certain to happen if we yield to the spritual disease I have described: if we allow ourselves to believe that any state or any class within a state exists for itself alone, and has no duties, no responsibilities towards the rest of the world. One alternative before us is mutual service and devotion, abnegation of self, of class, of race, nation, and language in the service of civilization and of the world; the other is to see Europe a desert, silent, unpeopled, uncultivated; riddled with the craters of shells and scorched black with the fumes of poisonous gases.

There is no third alternative.

May 7, 1919.

18

The Three Laws of Politics[1]

§1. *Community and Society*

A number of human beings living together constitute either a mere community or that special kind of community for which, when words are used correctly, the word 'society' is reserved. This is the Roman law-term *societas*; and *societas* is defined as a number of partners, who must be adult, free, male, Roman citizens acting together in the pursuit of aims advantageous to all, their will so to act being declared by each to the rest in a 'social contract', or mutual declaration to that effect.

In modern usage certain elements in this idea of society have disappeared, for example, the idea that partners must be Roman cititzens; but when allowance has been made for cases like this the word 'society', *société*, *società*, in European languages to-day still means[2] what *societas* meant to a Roman jurist: a community whose members are members of their own free will.

A number of human beings living together, whether of their own free will or not, I beg leave to call a community, thus making it clear that every society is a community but not every community a society; and also that I do not wish to become implicated in questions about non-human communities, the study of which, however attractive, is a blind alley for the student of human politics.

There is no living together unless something is shared; not as a family likeness may be shared, but as a family umbrella or a family hut may be shared; that is, by an arrangement binding upon all

Hobhouse Memorial Lectures, 1941–50 (Oxford University Press, London 1952). Reprinted with the kind permission of Oxford University Press.

[1] The author's thanks are due to the London School of Economics and Professor Carr-Saunders for arranging to deliver this lecture in his absence on the grounds of ill-health; also to the authorities of the Clarendon Press for permitting the lecture to be in substance an excerpt from an unfinished book, namely Chapter XXV of *The New Leviathan*.

[2] The word *Gesellschaft* is not included; that word is not derived from *societas* and not an equivalent for it. Quite commonly it means a non-social community; emphatically, for instance, in Marx.

members of the family as to the conditions under which it is at each one's disposal; the sum total of such arrangements forming what I call the *suum cuique* of the community. What makes any community a community is its having a *suum cuique*; what makes it a community of the kind it is, and not a different kind, is its having the *suum cuique* that it has, not a different one.

The *suum cuique* of a community may be due to a deliberate decision made in concert by all the members of the community, deciding (for example) what place in the communal hut each member of the community shall be entitled to regard as his sleeping place. To that extent and in respect of that activity, the activity of deciding where to settle down to sleep, the community is a society. In other respects it may nevertheless be a non-social community in which people, for example, sit down to meals in a decent and orderly manner for fear of a strict father. A community which in some respect is social and in others non-social I call a mixed community.

The mark distinguishing a society from a non-social community is that the members of the society act, in those matters which concern their membership of the society, of their own free will; members of the non-social community, in matters concerning that membership, under constraint. The essence of constraint is that a man should be brought into a condition of emotional disturbance in which he cannot make the free decision. The disturbance may be of many kinds; for example, it may be a state of fear.

§2. *The First Law of Politics*

A community engaged in political life is not quite a society and not quite a non-social community. It is a society so far as it is dealing with political problems by an act of the common will; a non-social community in so far as the same business of dealing with political problems is effected by a governmental action in which a part of its own weight is felt as a dead weight on the government. Of these two conditions a politically trained community tends to operate relatively often under the first; one which had to operate under the second would be either politically incompetent or else trying to deal with matters beyond its strength. At any point in the development of a body politic's experience, and therefore in any body politic, there will be in it a distinction between two classes of its members; the class which decides how to deal with a political problem, and the class which, with regard to a political problem now being dealt with, does what the first class tells it.

These two classes are called the rulers and the ruled. Such classes form themselves almost automatically in every body politic in relation to any political problem; and having formed themselves disappear as soon as their work is done; though there are psychological reasons for expecting a man or group of men whose advice has once proved acceptable to offer acceptable advice in the future.

That there is a distinction between rulers and ruled is the First Law of Politics. The distinction is always present in every body politic of every kind. No one even for a moment thinks that it is ever absent. When Louis XIV said '*l'état c'est moi*' what he was doing was to state that the ruling class in France (I return to the meaning of *état* below, §6) consisted in a corporation of one. When Abraham Lincoln spoke of 'government of the people, by the people, for the people', as characteristic of the United States, he did not imply that in that country there was no distinction between rulers and ruled; by 'people' he meant (what the word always means in political English, as opposed to the vulgar English which I here distinguish from the formal political vocabulary) *populus* as opposed to *plebs*; in speaking of the American people he did not intend to embrace under that designation new-born babies, any more than, as a matter of fact, he intended to include women. According to the tradition Lincoln was here expressing not an English tradition but a French tradition, the people of a country consists of its 'citizens'; and the citizens of a country are not its population but such of its population as have the rights which together make up citizenship.

If men lived for ever, the ruled class in a given body politic might, for all I know, disappear. Whether it would really disappear or not I am not interested in asking; for in fact men are so born, and destined to a long period of infancy. What is the legal status of a child that is too young as yet to exercise any political function? The ancient Greeks, with characteristic hardness of head, argued that if a child were a citizen it would behave like one; which it doesn't do, therefore it isn't one; it is a citizen's child, which is no more a citizen than a citizen's wife or dog is a citizen. The modern European thinks of the matter differently, never mind why; and give a citizen's child, at any rate, a citizen's son, the *prospective* title of citizen. For him, therefore, children (male children) of its members form a class of members of the body politic which are relegated to the ruled class on grounds of infancy alone.

§3. *The Second Law of Politics*

The First Law of Politics is concerned with the statics of the body politic; the Second Law with its dynamics. The First Law describes the body politic at any moment as divided into rulers and ruled, the Second Law describes this division as *permeable*. Those who rule or are ruled at any one time do not necessarily retain those functions; on the contrary, there are certain reasons on certain types of occasions for them to change to a different function; in particular to change over from being ruled to ruling.

With most of these reasons I shall not deal. But there is one which in a special sense operates constantly, namely, death. Different members of the community have different expectations of life; but the rate at which the expectation of life diminishes is constant. Take two members of the body politic; let one be a sickly specimen of the ruled class and the other a robust and energetic member of the rulers; whatever length of life each may expect, each expects one year less than he did this time last year. Men, as we know, are born babies; and a long infancy must elapse before one of them is fit for membership of the ruling class. Irrespective, therefore, of special hardships, dangers, or their opposites, to which members of one class may be specially liable as compared with members of the other, members of the ruling class start with a handicap. They are on the average older men than members of the ruled class; and their death-rate is proportionately higher.

There is, accordingly, a constant drain on the membership of a ruling class tending to depress it in numbers relatively to the ruled class; rulers die off. This drain continues to be felt however the ratios between the two classes are fixed. Suppose at a given time there are ten rulers to ten ruled, and suppose this is a convenient figure for the efficient working of a community. Then at any later time, say ten years later, death will have removed more of the rulers than of the ruled; not only because more of the rulers will have died, but because the ruled class will have been recruited by all the babies born during that period; unless, indeed, some of these turn out such prodigies as to be capable of ruling before they reach the age of eleven.

The result of these conditions will be the rapid extinction, utter or partial, of any governing class whatever; and to guard against this every body politic recognizes the Second Law of Politics; that the *membership of the ruling class has to be kept up to the required level by recruitment of competent persons from the ruled class.*

The Second Law of Politics specifies action to be taken by a community for the sake of avoiding the exhaustion of its ruling class and therefore its political extinction. What is thus specified is not a thing which will happen automatically; it is for the community to invent measures which are feasible in themselves and will secure the result aimed at. It is assumed not only that the body politic, and in particular its ruling class, sees the necessity of such measures and whatever they involve, but that the body politic as a whole is in a healthy enough condition for a lead in these matters to be followed.

No political law enforces itself automatically. It might seem at first sight as if the First Law did so; but it does not; granted a body politic confronted by a problem that calls for solution, it does not follow automatically that those members who are capable of solving it should solve it, and that the rest should follow their lead. In order that this should happen the body politic throughout its fabric must be sensitive to political realities, quick to recognize political leadership, and resolute to draw the consequences of that recognition.

§4. *The Third Law of Politics*

Whenever a body politic is active, that is to say whenever any form of political activity is going on in it, there is always a nucleus at which the activity is concentrated; the activity, taking its rise in a certain part of the body politic, communicates itself to the rest of the body in various ways and according to various principles. The nucleus of the activity may be described as having the *initiative* with regard to this event; the distinction between the part which has the initiative and the rest has been already described by the First Law of Politics as the distinction between those members who exercise rule and those over whom it is exercised.

It will be useful here to recall a technical term of Greek political theory. I refer to the Greek word ἀϱχή, which represents a simpler idea than the word 'rule' by which we commonly translate it. The Greek conception of the activity which we call ruling is the conception of opening new paths in thought or action or what-not, paths which are thereafter open to other men. According to the Greek idea of him a ruler is a path-finder, a pioneer; and typical examples lie ready to hand in the Seven Sages of Greek tradition.

Translation is a dangerous game; and to translate the Greek phrase 'Seven Sages' is to represent it by something that misrepresents the spirit of it. Sages in English have long beards and more than a hint of second childhood. Perhaps Greeks wore their beards with a difference; at any rate a man who turned to a friend in the heat of battle saying 'Mark you, I would not be elsewhere for thousands,' has to the English imagination the clean-shaven face of youth; but to the Greek mind he is a Sage; a wise man; a man who knows how to win battles.

The qualities which a ruler must exhibit in his demeanour (and his 'demeanour' is by no means a negligible part of his activities; it means activities publicly performed in the eyes of what is called 'the world', namely, those whom the ruler has to rule) are of the most varying kinds according to differences in the problem he had set himself to solve. Whatever qualities are thus exhibited in the course of ruling are exhibited as models for imitation. The ruler as path-finder is the ruler as setter of examples.

Here we have a principle which I call the Third Law of Politics. Into the psychology of it I will not enter; nobody, I think, will deny that the position occupied by a ruler is such that the characteristics displayed by him in the course of his activity of ruling will in fact tend to be imitated by those over whom he rules; that the tendency will be stronger in proportion as the bond between ruler and subject is closer; and that the fact of this imitation, which in any case will to some extent occur even unconsciously, will be replaced in proportion as the ruler becomes master of his trade by his deliberately offering examples intended to be worth following.

§5. *Digression 1: Bodies Which are not Quite Political*

A famous lecturer on politics, T. H. Green, refused to Russia the name of a 'State' because it had too strong an inclination towards despotism for a name to suit it,[*] part of whose connotation was

---

[*] T. H. Green contends that: 'We can only count Russia a state by a sort of courtesy on the supposition that the power of the Czar, though subject to no constitutional control, is so far exercised in accordance with a recognised tradition of what the public good requires as to be on the whole a sustainer of rights.' *Lectures on the Principles of Political Obligation*, with a preface by B. Bosanquet (London, Longmans, Green, 1917), 137.

freedom. Green has been held up to criticism for saying this by so eminent a political theorist as Professor Laski,[†] in a very recent Hobhouse Lecture. Let me briefly, in my own words, indicate my agreement with Professor Laski.

Green, if I understand him, was allowing the issue to become confused in his own mind. Unless I am mistaken, he was not clear as to the nature of the act which he was expressing, or was about to express, or had expressed, by uttering the words: 'Russia is something less than a body politic,' or, in his words, 'a State'. Such a formula, as uttered by a speaker in Green's position, is likely to be ambiguous.

(*a*) The ostensible or obvious meaning of the phrase is *scientific*; and in particular classificatory. It means: 'Here we have something whose superficial characteristics would lead you to classify it as an *x*. Do not be deceived; look again; and you will see that it is really a *y*.'

(*b*) Very closely connected with this is a *practical* meaning of the same phrase. It runs: 'Here we have a thing towards which we might act in either of two ways. Now is the time to make up our minds between them. Do not be seduced into treating the thing as an *x*, treat it as a *y* and damn the consequences.'

I hope I need not pause to argue, either that these two ideas are different, or that they are easily confused owing to the identity of the phrases that commonly serve us to express them; if there are any among you who would like to see me argue these questions I beg to be excused.

What Green was doing was to utter a statement about the contemporary character of Russia, which was in point of fact a practical statement, as if it had been a scientific statement. In effect what he said was that the Russia of his time had no business in the civilized world and that decent people would not recognize Russians as equals. This is as if someone had asked a second party what sort of meat had been offered him, and the reply had been given, 'It is a perfectly beastly sort of meat'; telling you not what the meat it, but what his practical reaction to it is.

Green was a man of strong political views; as a citizen of a

[†] H. J. Laski, 'The Decline of Liberalism' (Oxford, Oxford University Press, 1940). Reprinted in *Hobhouse Memorial Lectures 1930–1940* (Oxford, Oxford University Press, 1948), 24. I can find no reference to Green's criticism of Russia in this lecture.

university town he felt his duties, including his duties as a Liberal, very keenly. Russia to him was a country where men of his kind were persecuted, banished, imprisoned, and in a word treated very much as in our times their likes have been treated by Nazi and Fascist governments. What Green did was to attach the infamy of these facts indelibly to the name of the Tsarist government. His indignation got the better of his scientific bent.

Should not a cobbler stick to his last? Was it not Green's business as a political scientist to analyse facts without indulging in condemnation? It is a question that has to be answered with some care, for it is one thing to assume the garb, whatever it is, of a scientist and another to assume that of a judge. It is a really elevating spectacle, Green scolding the Tsar of all the Russias, and by anticipation Hitler and Mussolini into the bargain; or is it a more elevating spectacle, Spinoza making and obeying an iron rule against condemning anything that the human mind can offer to his gaze? (Spinoza, Appendix to the First Book of the *Ethics*.)

Cast your minds back, gentlemen, to a little before the outbreak of the present war, when the horrors of the concentration camp were familiar facts to all of us, and we know that they were at least equally familiar to members of the Chamberlain Government, who were concealing them from the general public to prevent indignation from flaring up into an inopportune and hopeless war. Was this deception a crime against the public or not? It is an expensive luxury to indulge a disapproving attitude towards the internal affairs of a country; not only an expensive luxury, but likely to prove disastrous except for a people that knows what kind of debt it is running up and stands prepared to pay the bill.

Freedom is certainly part of the connotation of the name 'body politic'. But not the whole if it. That is what is wrong with Liberalism as a political doctrine. For part of the connotation is the opposite of freedom; as everyone knows whose study of politics has brought him as far as the First Law. There is always in any body politic a ruled element whose function in the body is to accept the initiative of an element stronger and more active than itself.

§6. *Digression 2: The Use of the Word 'State'*

T. H. Green used the word 'State' as a synonym of what I call a body politic. I crave your patience to extend my digression for a moment and to consider this use of the word. Science consists in using words scientifically; when words are used unscientifically there is no science.

Lo Stato, état, estate, is a technical term in medieval politics for what we, having followed Marx's lead at this point, have decided to call a class. Among the classes or estates which together make up a body politic, medieval thinkers recognized one which enjoyed a monopoly of political power and glory. By a not intolerable figure of speech, since this class did all the ruling, for political purposes you might refer solely to this class in discussing the body politic. Thus *Lo Stato*, the State, is occasionally used in the fourteenth century by synecdoche for the entire body politic.

This use of the word had only a short life, unless we take into account its long life as a ghost, to which I shall come in a minute. It became obsolete as soon as political science recognized what I will call the *positive function of the ruled* in the life of a body politic. To be a mere recipient of a ruler's behests, an obedient subject, is to have a merely negative function; to have a positive function is to have a will of your own which your ruler must take into account. The chief lesson which Machiavelli learned from his famous study of the fortunes of Louis XII on his Italian campaign, and set forth in the third chapter of *The Prince*: 'Concerning Mixed Principalities', concerns the way in which a prince may use his subjects as a reservoir of strength; they thus become no longer negative or passive partners in the work of government but active participants in it. As soon as Machiavelli had made this discovery the use of·the word 'State' for 'body politic' became obsolete. I confess it with shame; I had thought that Machiavelli had lent his authority to the survival of the obsolete term, but I find that he did not; the word 'State' is of common occurrence in the text of *The Prince*, but in the cases I have examined it means, as in Machiavelli's mouth it ought to mean, not a body politic but a ruling class.

The word 'State' as a designation of a body politic had thus disappeared from good Italian by about 1513. The fashion came back into use among the Germans of the late eighteenth century; I do not know why, though I suspect that it may have been due to a hasty and ill-informed imitation of Renaissance Italian. I notice that the habit is absent from the earlier and more scholarly writers such as Leibnitz; and outside Germany is found only where German usage has been uncritically copied, as I am afraid it was by T. H. Green and a circle of other writers who thought a German book a sufficient authority and its author's nationality all the guarantee of which his words stood in need. Outside this circle of German and germanizing writers, the word 'State' is never to my knowledge used as an equivalent for 'body politic'; and the quotations of supposedly confirmatory Latin extracts given under *Staat* in Krug's *Philosophical Lexicon* are, I suspect, all as baseless as those of them which I have taken the trouble to test.

I put it before you as a suggestion which rather shocks myself, perhaps because of Burke's phrase about bringing an indictment against a nation, that the so-called *theory of the State* which has occupied so much time during the nineteenth century more than deserved the obloquy all too timidly heaped upon it by writers like Duguit[‡] in the twentieth; that it is due to a German misunderstanding of the whole European tradition of politics; and that this misunderstanding has been deeply rooted in Germany for hundreds of years. I would gladly stop to defend this suggestion and to show how deeply rooted in Germany[3] is the incapacity to understand the elements of political theory; but this is only a digression, and it is time it came to an end.

§7. *The Third Law of Politics, continued*

I have mentioned the ἀρχή which distinguishes the rulers from the ruled in a body politic. The conception is a purely relative one. It is relatively to the ruled that the ruler had initiative, command, or whatever we call it; but what is it in itself? It cannot be power; for the ruler's power is his power *over his subject*. It cannot be intelligence; it is not his intelligence as such, but his *superiority in intelligence to his subjects*. The ἀρχή which distinguishes the ruler is nothing at all in itself. It is only some currency in which he is richer. Any currency will do; but it must be accepted on all hands as one that does pass current. It must be something in which the ruler is accepted as superior to the ruled. ·

---

[3] I will illustrate this incapacity by one quotation. 'It is a great matter to live in obedience; to be under a superior, and not to be at our own disposing.' Thomas à Kempis, a popular devotional writer, in *The Imitation of Christ*, chap. *ix* (C.1400).

[‡] L. Duguit, *Law in the Modern State*, trans. F. and H. Laski (London, Allen and Unwin, 1921).

Among the blind, says the proverb, the one-eyed man is king. Long ago Mr H. G. Wells wrote a fantastically unpleasant story pointing out that not this, but the opposite, might turn out the case, men being as touchy as they are about scales of value. He depicted a body politic whose life was so attuned to blindness that any departure from that standard was more than repugnant, it was criminal. Now the idea of intelligence as conferring upon men power to rule is due to the legacy we have inherited from the Greeks, who have presented us with the portrait of a rational or intelligent man, and have shown how a number of such men would build up a corporate scale of values, current in their own community, in which intelligence or rationality is that which makes one man suitable by common consent to command another.

I do not say that this Greek scale of values is capable of being systematically reversed; or even of being systematically replaced by some rival conception of human excellence. What I say is that the Greek scale may suffer interference, and has on certain occasions suffered a certain degree of interference, from a different scale or scales. Examples are not far to seek; let me ask you to reflect upon one which never fails to impress me because of the contrast it involves between what some would call a 'classical' or Greek background and a foreground of a disconcertingly different kind. Think of the mad Roman emperors of the Julio-Claudian line, and suppose (if you can) that their story has been more or less faithfully transmitted to us; suppose, I say, that the famous disclaimer of guile or spite on the part of Tacitus is, incredible though it may seem, correct; and that the things which he describes in that nightmare narrative really happened. Imagine that Nero, for example, was as mad as they say he was; but that there was a kind of method in his madness which enabled him after a fashion to rule a great empire. When I speak of a method in his madness I am not indulging in a meaningless quotation of *Hamlet*; I am not speaking of a madness that operates only according to the direction of the wind, so that the victim is mad only in certain respects or at certain times; I speak of a personality completely corrupted by mania, doing and saying only what the mania prescribes.

You may be inclined to doubt whether the suggestion I am going to make is sense or nonsense. However, it is in Plato; it is that third category of which Plato talks, midway between being and nothing;

it is in the Fathers and Doctors of the Church, and a well-brought-up Catholic or Orthodox Christian will find it familiar and will, perhaps, cross himself in the belief, perhaps not mistaken, that I am talking about devils. A student of modern psychology would agree in finding it familiar; but unlike the orthodox Christian he would not use a ritual gesture to defend himself against what was being talked about, he would prefer to use ritual vocables; he would not call upon the name of God, he would prefer to speak about the inventions of Freud, calling them complexes and I know not what.

What I suggest is that, whether or no this is recognized by the accepted or Greek theory of human intelligence, there are *two kinds of unintelligence* in the world, and that these have different functions. There is what may be called a *negative unintelligence*, which is the thing of which Plato says that its proper object is nothing at all; a person in this frame of mind, trying to grasp something, grasps nothing; he comes away from all mental effort empty-handed. The other kind of unintelligence is a *creative unintelligence*, creative of chimeras and nightmares; unintelligence of this kind creates these things more profusely according to its own fecundity; this fecundity being a positive power in so far as it creates, but a mere absence of power in so far as what it creates is nothing at all. The world is in no sense the richer or fuller for all its creative efforts. And in this sense it is all one whether you talk about this positive unintelligence or that other negative unintelligence which I mentioned first; in either case there is nothing.

The question with which we are dealing is this: how can a man, without being intelligent, acquire that mastery over men which the Greek theory of life ascribes to intelligence?

We cannot say that the Greek theory is simply wrong, and that intelligence has no such initiative. Proof to the contrary is forthcoming in abundance for every man in the experience of every day. But if unintelligence were a purely negative thing, a mere failure on the part of the intellect to grasp its object, unintelligence could not thus simulate the fruits of intelligence.

The answer is that there are *two ways of being a fool*: you may be *foolish to stupidity*, so that your mental hands grasp nothing of what they try to grasp; or you may be *foolish to craziness*, so that your mind creates illusions or hallucinations about the things of which you are trying to think. These two kinds of foolishness occur in practice much confused together. The stupid fool, in politics as elsewhere, creates nothing; the crazy fool creates much; although this much, being crazy, comes to nothing.

But in the meantime, not yet having been weighed, the crazy fool presents us with the aspect of being a formidable producer. This is in general terms the explanation of things like Nero, of which Tacitus and the whole of Roman history had not a word to give by way of explanation. Small blame to Tacitus; even the greatest brain of Greece had not gone deeply enough into the theory of error to offer him the blue-print of a solution. Plato had an inkling of the truth; but not more than an inkling; Aristotle not even that.

The crazy type of fool can pretend to be wise. The fertility of his diseased mind gives him an initiative, futile it is true, over his fellow men. He has just as much initiative as a man who is really intelligent; in one sense even more, for he has less to fear. The intelligent man offers himself to an equal wrestling bout of minds; he stands up to all comers, and faces criticism; he does not know from what side criticism is going to come, or that it will not prove him to have made a mistake. The crazy type of fool with his psychological hold over his audience will easily convict him of being a fraud; which, strange though it may appear, is rather a feather in his cap than a thing to be ashamed of.[4]

§8. *The Platonic Tyrant*

Plato, in the ninth book of the *Republic*, has given his readers a memorable description of what he calls a *tyrant*. By a tyrant he does not mean what we call a despot, or ruler who rules for personal motives and with considerable display of cruelty, arrogance, and other qualities valuable to him chiefly in their enhancement of his personality. The despot, with all this emphasis on his personality, may have something to emphasize; the laws which he administers with cruelty may be wise laws and justly administered. There may be a barbaric swagger about him, but it may serve to lend *éclat* to a genuine political performance.

The tyrant, on the other hand, puts up no political performance. He is merely so much jetsam, floating on the surface of the waves he pretends to control. His qualities, according to Plato's scale of values, are not the qualities of a free man, let alone those which

---

[4] A thing I noticed in Italy in 1939.

would enable him to be a ruler of free men, but the qualities of a slave. He is not the sort of man who can triumph over his own weaknesses; more like the sort of man who would yield to them on every occasion; his progress through the world is a rake's progress supported by burglary, pocket-picking, and other low forms of predatory activity, preparing the way, says Plato, for higher forms of thieving such as robbing temples; or, as we should say, confiscating deposits in banks. His rise to the position of tyrant is consequent on a class movement; it is concurrent with a rise of the lowest social class in the city to the position of gangsters patronized by himself; it is not his own strength or energy that lifts him to the position of tyrant but, so to speak, his low specific gravity. It is in his capacity as so much jetsam that he rides effortless over the waves of politics.

In this passage Plato is describing something he has seen going on in the world around him. His description has a sharpness and a liveliness not otherwise to be explained. If all other records of the Hellenistic age were blotted out, we should know it on Plato's word for an age of tyrants; we should know that the human animal Plato has so vividly described was a kind of fauna produced by the special conditions of that age. The word 'tyrant' is not always used by Greek writers with scientific accuracy. If its right sense has been determined once for all in this passage, the application of it to the absolute rulers of the age of Pisistratus is not quite correct. I will not spend time on the point here.

It is more important to notice that the political phenomenon first noticed by Plato has a long history before it. From Hellenistic Greece it infected Imperial Rome and produced the Julio-Claudian mad emperors; and thereafter it lingered in shapes I will not attempt to describe as an infection from which certain parts of Europe have never been entirely free; one might describe it, since it is essentially a phenomenon of declining Greco-Roman life, as *Renaissance sickness*; the sufferers have the illusion of a healthy and vigorous youth when they show the symptoms of an embittered and destructive old age; the most distinguished of them at the moment are Mussolini and Hitler.[5]

The phenomena of tyranny were a disease to which the Greek type of political experience was fatally liable. Reconstruct that type of political experience (a thing deliberately done by movements

[5] The first, now, only in retrospect (1941).

professing to restore the 'antique') and that disease will reappear. That is why I called it Renaissance sickness. Owing to the success of Christianity in constructing a political experience of a new type, the disease was for many centuries kept at bay after its first epidemics in the pagan world.

§9. *The Reversed Action of the Third Law of Politics*

The disease works by what I will call a reversed action of the Third Law of Politics. Like every other political law, this one does not enforce itself automatically; men must take trouble to obey it. Its direct action begins with a body politic composed of what we call sane men; the result is that they accept the leadership of sane men; and the leadership of the sane produces, with unremitting labour, a rank and file of sane men. Where, you may ask, does all this labour go to, all this running to keep in the same place? The answer is: it is the work done by the community in keeping itself sane. It is much easier for any kind of man known to me to doze off into daydreams which are the first and most seemingly innocent stage of craziness. If labour-saving is what you want, give up all this trouble about thinking; go mad and have done with it. That is what the tyrant has to offer mankind—an end to the intolerable weariness of sanity.

The reversed action of the Third Law of Politics is precisely this cessation, on the part of a body politic, of the effort after sanity. The engine has slipped into reverse; and the whole thing, with delicious absence of exertion, is sliding downhill. It is much easier to speak and act and write crazily than to do it intelligently; you just let yourself go, and there you are. This is the first phase of the reversed action. The next phase is that the resulting 'democracy' (as Plato and Hitler, strangely united for once, agree in calling it) creates leaders for itself, leaders from its own members, leaders of fashion in the temporary freaks of craziness, under whose tyranny the whole body politic lets itself go more completely than ever; for to shout with the mob (that is, to obey the tyrant of the moment) is the easiest thing anybody can do.

§10. *Conclusion*

The rise of the tyrant is just as natural, just as normal, as anything else that happens in the political world. If what is natural or normal means what is recognized by the laws of any science as happening according to those laws, it is as natural or normal as anything else; if it means easy or effortless, it is a great deal more so, as letting your ship go to the bottom is easier than making any kind of effort to save her. For it proceeds according to the laws I have laid down; the only difference is that it involves the saving of the labour which, in what we have learned to call ordinary life, is spent in the constant generation of sanity; and what a saving that is no one knows who does not keep himself sane by hard work, and also watch himself with a practised psychologist's eye to see what labour it involves.

Once someone has somehow hit on the brilliant idea of saving all that labour, the modern tyrant has embarked on his career; which in the conditions of the twentieth century is certain to be epidemic. Those who are trying to resist the infection are doing nothing but keep hold upon their sanity; those who are engaged in spreading it are urged on to their missionary labours by a kind of benevolence, a desire to liberate those who still groan beneath the burden of sanity. Where resistance against the infection is unsuccessful, those who so describe it may have misunderstood the nature of the disease; it may be that a given body politic was immune, not against that disease, but against a different one.

To many of us, myself included, the collapse of France before Hitler a year ago was a shock. I knew from experience of the country how stoutly the French were armed against what they called '*les sales Boches*'. I did not believe that they would ever accept them as conquerors. Now that they have been conquered, I ask: by what? Such a conquest must have been due to some poison which the French blood had not secured itself against; not, therefore, the poison of Pangermanism, against which it was immune.

What can this have been? We have made very good friends with the French since Waterloo; but there has always been a sore spot in our friendship. We were the people who beat Napoleon, and the French never wholly forgave us; the *légende napoléonne*, scotched but never killed, the nostalgia that possesses every true French heart for the crazy Corsican, the little fat man with a star, which once already, since Napoleon's death, has led France to flirt with tyranny, has revived in force, and urged her into a policy at which you can see the ghost of Nelson smile grimly as one who knew it all the time.

I have given you a brief survey of the fundamental laws of politics, designed to be altogether neutral as between any one system of politics known to me and any other; and equally adapted to explain any and every such system. I have especially tried to show how they will serve to explain the political crises and disasters, as from our point of view they are, under which we now suffer; which, I am persuaded, a properly thought-out system of political principles can explain as easily as a happy political life. Politics, like other sciences, if it can explain the good can equally explain the bad; and I will add that the way to explain them both is in the first instance to say nothing of either.

Neither these laws nor any other, if they are scientific laws, can tell us who is going to win this war. Questions like that cannot be asked of scientific research; they are properly asked (if properly is the right word) of fortune-tellers or other dealers in the occult. There have been German philosophers, I am thinking especially of Hegel and his follower Marx, who believed that the future in its general lines admitted of being scientifically foretold; more fools they. Our relation to the future is not that the future, while it is still future, is to be foreknown by us; the future can be known only when it has become the present; but that it has to be made by us, by the strength of our hands and the stoutness of our hearts.

19
Draft Preface to The New Leviathan

The most conspicuous thing going on in the world today is the revolt against civilization.

What civilization is and entails I shall try to say in another place. Here I will only say that it includes these three things: law and order, prosperity, and peace.

It means a condition of society in which men live according to definite rules; these rules being known to all, so that no one either obeys or breaks one of them without knowing it, and enforced, by men to whom that duty is assigned, by penalties imposed upon any one who breaks them. A society whose common way of life is thus defined by rules and sanctioned by enforcement of these rules is said to be in a state of law and order.

It means a condition of society in which a man acquires the things he needs for his sustenance and comfort not by taking them from other men but by 'winning' them in some way which does not deprive others of them; either by taking possession of them as *bona vacantia*, or making them for himself, or purchasing them from someone who owns them but does not need them, in exchange for something that he himself owns but does not need. A condition of society in which every member of it pays his own way by 'winning' his livelihood, instead of obtaining it parasitically by the impoverishment of others, is (so far as the natural and technical resources of that society permit) a condition of prosperity.

It means a condition of society in which the arbiter of human affairs, whether as between man and man or as between one group of men and another, is not violence but agreement; and this is peace. This condition includes the other two. In its legal and political aspects, peace means law and order: the supersession of violence, as between man and man or group and group, by 'orderly' conduct, that is, conduct which is not violent but proceeds according to rules or laws agreed upon by the parties concerned. In its economic aspect peace means prosperity: the supersession of plunder-economy,

'NL Fasc. 1.2. = Preface', Collingwood MS DEP 24, pp. 3–9.

which enriches one man through the impoverishment of another, by production-economy, in which every member of society enriches himself by 'winning' goods which, unless he won them, would either be of no use to anyone or else would not exist at all.

The revolt against civilization is based upon two accusations brought against civilized society by the rebels.

First, that it is fraudulent. The society which calls itself civilized, we are told, does not live up to its own ideal. It preaches peace, but it makes war: imperialistic war against other societies, class war of the exploiters against the exploited within its own body. It proclaims law and order, but its laws are only a cloak of legality thrown over its own oppression of its weaker neighbours and the oppression, within itself, of the weak by the strong. Its order is only a device for depriving the oppressed of means and will to resist the oppressors. It pretends prosperity, but it lives in the most abject poverty. It possesses enormous powers of creating wealth, but this power is not used. The persons who do productive work, in the communities which call themselves civilized, are not allowed to produce all that they could produce, for fear the market should be spoilt by over-production; and the value of what they so produce goes not into their pockets but into the pockets of the men who are exploiting them.

Secondly, that it is misdirected. The very ideal of civilization is false. A civilized man is a bad man, a bad specimen of the human species. The qualities he achieves in so far as he is civilized are vicious qualities. A civilized man is not only a coward, but religious in it. His chief deity, peace, means nothing but his own safety from danger and hard knocks and all the things that frighten him: blood, and fists, and angry eyes. Of his two minor deities, law and order means the principle that someone else, who is paid to do it, shall stand between him and the things that frighten him; prosperity means soft living, flabby muscles, the full belly, and the bursting bank-balance.

These accusations are today the official doctrine of a country which is at war with our own, and of the ruling parties in at least three other countries. This already creates a situation demanding a good deal of thought. The demand is further strengthened by the fact that the same accusations have impressed themselves deeply for a long time past, but with increasing urgency of late, upon the consciences of a great many persons all over the world.

Here in England we are at war with the country which has taken the lead in the revolt against civilization. We know what our enemies are fighting for. They are fighting for the destruction of civilization. A successful war, for them, does not necessarily mean a war in which civilization is altogether destroyed. It is enough that civilization should be signally weakened, its forces signally diminished, the faith in its value signally impaired.

What we are fighting for, nobody knows. The official doctrine of our government is that we are fighting to defeat the enemy. In a sense, that is true. All fighting is directed to the defeat of the enemy. That is what fighting is. In the same way, all travelling is going on. But if you asked a man who was guiding you on a journey 'Where are we going?' and he said 'On', you would infer either that he wished to conceal the destination from you, or that he did not know it himself.

At risk of stumbling upon truths which are meant to be kept hidden, therefore, I have tried to find my own answer to the question. Our enemies are fighting in prosecution of the revolt against civilization. If we really wish to defeat them, I suppose we are fighting in defence of civilization. But among ourselves, as I know very well, there are many who doubt whether civilization is worth defending. And I do not know how these doubts can be resolved unless we first make up our minds what civilization is. I have already defined it as a condition of society involving law and order, prosperity, and peace. But I have only indicated in a very rough and vague way what those three terms signify, and not at all why the things they signify should be thought desirable. I have implied that they are things which, if they are to exist at all, must exist in a society; but I have not even asked what a society is. We all know, I suppose, that a society consists of human beings; but I have not asked what a human being is.

Yet if we do not know what man is we cannot know what society is; if we do not know what society is we cannot know what civilization is; if we do not know what civilization is we cannot know whether the accusations made against it by those who are in revolt against it are justified or not; nor granted that they are justified, what they prove.

These are the questions I propose to ask in this book. I shall begin by asking what man is. Next, I shall ask what society is. Next, I shall ask what civilization is. Then I shall consider the revolt against civilization: and lastly I shall ask how a society which considers itself civilized should behave in the face of this revolt. It is the last of these five questions that constitutes the real subject of the book; but it cannot be answered until the way has been prepared by answering the other four.

I can imagine the reader saying to himself, as he closes the book at this point and replaces it upon the bookseller's counter, 'What is man? What is society? Pretty little questions to ask as a mere preliminary to a question in practical politics! The author must know that these questions have perplexed the wisest men ever since men first began to think. He must know that he is not going to spend on them more than a few thousand words. Does he mean me to believe that he thinks himself able in these pages to give any answers to them that it will be worth my while to read?' That is a fair criticism. I will answer it fairly.

Suppose you were not very learned about electricity; and suppose you were driving through a shallow river, in the middle of which your car stopped. Suppose you looked into the bonnet and found everything splashed with water; and concluded that water had put the ignition-system out of action. Suppose, moreover, that the river was rapidly rising in flood, so that in half an hour the car would be bodily washed away; and finally, that your passenger was an electrical engineer. Perhaps you would call out to him, 'Come here and show me how this thing works.'

If he were the most pedantic kind of imbecile, he might reply: 'We have only half an hour, so you must excuse me. It takes me sixteen hours to expound to my classes even the ordinary textbook theory of electricity, and even then there is the question whether the ordinary textbook theory is true.'

If he were a sensible man, he would understand that when you asked 'how this thing works' you only wanted to know just so much about 'how it works' as would enable you to cope with the present emergency. You wanted him to say something like this: 'See that thing? It ought to be dry. Got a rag? No, not that one, it's wet. Not that one, it's oily. Yes, handkerchief will do. Wipe it carefully; if you don't get it dry, there will still be a short. That is right. Now try starting her up.'

The purpose of this book is to ask one question and only one: How should a society which regards itself as civilized react to the revolt against civilization which is now going on? Unless that question is answered pretty soon it need not be answered at all, because there will no longer be any societies in that position. It cannot be answered, as I have said, unless certain other questions are answered first; but, granted the present emergency, what has to be said about these other questions is just what the present emergency demands, and no more.

As a professional man, the writer of this book is under an obligation to know a good deal about the answers that have been given to these questions. But he is under no obligation to display his knowledge here. In fact, to display it would be like lecturing one's patient on the circulation of the blood while he bled to death. What is contained in Books I–IV of this volume is, therefore, the barest minimum which must be known by every member of a 'civilized' country, whatever his profession or occupation, if in the present emergency he is to do his duty as a citizen.

A great many of these things have been already said by Thomas Hobbes in his *Leviathan*: so many that the present writer has stolen Hobbes's title and dedicated his book to Hobbes's memory. The *Leviathan* was the first book in which the idea of a civilized society was consciously and systematically expounded. The obloquy that has been heaped upon it for century after century is sufficient proof that a revolt against the idea has been long brewing. Now that the revolt has come out into the open, it is time the idea of a civilized society was expounded again, with what modifications the advancement of knowledge between Hobbes's day and our own may afford: and with this design in view, hoping, if he cannot earn any share of Hobbes's glory, at least to earn a share of the infamy that Hobbes earned by telling truth, the writer has ventured to claim for his book the title of

THE NEW LEVIATHAN.[1]

[1] The published 'Preface' is much shorter, and the fifth part of the book, which Collingwood described as the most important, was never written. Ill health may have prevented its composition, but also the purpose of the fifth part, to ask how society can face-up to the revolt against civilization, is partially fulfilled in the four parts that eventually comprised *The New Leviathan*.

Appendix One

After presenting his theory of question and answer in *An Autobiography* Collingwood confesses that he had expounded the theory at length, with numerous applications and illustrations, during 1917 in a book entitled *Truth and Contradiction*. He maintains that he submitted the book to a publisher who declined to take it up on the ground that the times were not right for a publication of that kind.[1] Collingwood tells us that he destroyed the only copy after he completed *An Autobiography*.[2] One chapter, however, remains extant and is among the Collingwood papers in the Bodleian library, Oxford University. In the chapter he subjects the ideas of coherence and contradiction to rigorous analysis[3] from the perspective of a theory which is not unlike the principles of concrete affirmation and concrete negation presented in chapter 5 of *An Essay on Philosophical Method*.

Collingwood offered *Truth and Contradiction* to Macmillan in a letter dated 8 January 1918, in which he describes it as about 60–70,000 words, of a mainly philosophical character, but also dealing, among other things, with politics, war, and relations between the classes.[4] Macmillan sent the manuscript out to Henry Jones for his advice. Jones, after voicing a number of reservations, recommended publication. Jones admirably captures Collingwood's character and the report is reproduced here both for its curiosity value and because it remains the only independent description we have of what Collingwood was trying to do in *Truth and Contradiction*.[5]

---

[1] Collingwood, *Autobiography*, p. 42. [2] Ibid. 99.

[3] 'Truth and Contradiction, chapter 2', Collingwood MS, DEP 16, pp. 1–21.

[4] The contents of the letter are described in Johnson, 'Letters of R. G. Collingwood', p. 25. Unfortunately, for copyright reasons, Collingwood's correspondence with Macmillan is not included in the microfilm edition of the Macmillan Archives produced by Chadwyck-Healey Ltd.

[5] Macmillan Archives, 3rd series, vol. mcciii, 1911–21, fos. 114–15. British Library Additional MS 55988. Chadwyck-Healey microfilm edition, rel. 7. Reproduced with the kind permission of Mrs Jean Hunt, Macmillan Publishers, and the the British Library which own the MS.

Henry Jones's report on R. G. Collingwood, *Truth and Contradiction*.
Noddfa, Tighnabruaich, Argyllshire.
4th March 1918
My dear Macmillan,

I have found Collingwood's 'Truth and Contradiction: A Study in the Development of Thought', an uncommonly difficult 'proposition', if I may use that Americanism. I have read every word of it, and done so with lively interest; which is as good a testimonial as I could give to a book. But I cannot feel that I have a clear estimate of its worth. Not that it is unintelligible, nor that its problems are unfamiliar, or its doctrine strange; but that it has such contradictory qualities.

I do not know any writer more frank. He cares not one whit to what extent he exposes his flanks to his critics, and makes statements which, taken by themselves, look either purely absurd or preposterously untrue. But that is only one side: on the other is the fact that these statements are *stages* or *steps* in the development of his main argument, half truths or sheer errors in which it is not possible to rest and which just compel a movement onwards to a wider truth. And this is precisely what he means, and rightly means, by 'dialectic', the way in which 'the development of thought' takes place.

Again, much of the book is sheer logical argumentation, acute, incisive, and merciless: the abstract play it looks of sheer 'intellect', in the bad sense which, unfortunately, that word is acquiring. His examples on these occasions look trivial, obvious, the smaller things of the mere logician. Then will come passages that are 'full-blooded', the pulse of a wide life beating there and the strength of the mature reflective thinker. He writes at times like a smart young don, and, at other times, as a Sage: and he can be quite crude.

The plan of his book is obvious, and good. He exposes the inadequacy of the theories of truth—'the Coherence' theory and 'the *Doctrine* of Degrees of Truth.' (And in passing I may say that his exposure of the former is more effective than his exposure of the latter). But I cannot say that the plan is thoroughly worked out.

By exposing the inadequacy of the other doctrines he means (and exemplifies) the discovery of the *truth* that the theories contain, and reconciling the half-truth each contains in a wider truth. This is a version of the 'dialectic' of Plato and of Hegel; and in a sense not new. But it is done in a fresh way: clear, frank, interesting, and some-how very 'taking'.

Having exhibited the dialectical movement in 'thought', he reveals the same movement in Art (more particularly in *Music*) and in Morality. And this part is fresh.

All the time he is showing the true nature of Philosophy, and finding that *movement, activity, process* is the living soul of all thinking and of all objects of thought. Having reached his goal I wish he expounded it more fully.

Yes! I would publish the book in spite of its defects: The author will 'get beans' for his frank, little tutorial indiscretions, for he is a college don from head to toe, I'd say: but he has something real to say. I cannot but like the little book, with all its faults: for it *is* as far as it goes a 'Contribution' and step forward.

Yours sincerely,
Henry Jones

PS I return the MS registered as usual.

Appendix Two

Among the papers of Sir T. M. Knox, held in the Library of St Andrews University, Scotland, are to be found various items relating to the Collingwood family. The extracts from the letters of R. G. Collingwood to Malcolm Knox reproduced below are of direct relevance to the subject-matter of this book. I would like to thank James Connelly and Peter Johnson for alerting me to the existence of the letters, and Mrs W. G. Morton for granting permission to publish them.

Letter to T. M. Knox from R. G. Collingwood. Knox manuscripts, University of St Andrews, MS 37524/421. Dated 2 Nov. 1937.

There is a pestilence abroad, and its symptoms in withered minds and paralysed wills are all round us. Most of our colleagues are its enthusiastic allies or its helpless victims, and I don't believe that controversial argument is the way to attack it. When I was a beekeeper and the acarine disease was killing off all the bees in England, I kept it at bay by constant re-queening, and thus pouring out constant floods of new blood faster than the disease could infect it. I think that is the way to deal with the pseudo-philosophies of today. Go on producing good stuff—not negative or controversial stuff, but meaty nourishing stuff, and drive them out of the field by showing that we can appeal over their heads to the people who need philosophy and will not be content with the sophisms of our friends.

After which: here is a question. Is there really such a thing as *moral* philosophy? I mean, is there a philosophical science of action as distinct from one of thought? I am pursued by the notion that these are not distinct or different at all, but that there is only one thing which in your paper you call (and call well) Action, for which thought would be an equally appropriate name. I'm thinking, I suppose, of the kind of position taken up by that great man Lachelier in his contention that knowledge is the same as freedom. And I raised the question because I don't know how far you would go in your repudiation, in the first paragraph, of the dualism between 'theory' and 'practice'. I would go a terrible long way myself.

Letter to T. M. Knox from R. G. Collingwood. Knox manuscripts University of St Andrews, MS 37524/429. Dated 3 Sept. 1939

. . . If Russia is not brought into the war against us, we shall win. And it will not, I believe, take very long. Both we and France are much stronger than in 1914; Germany I believe is weaker. The Germans are already hungry. Their nerve is strained by the Nazi regime into a dangerous brittleness. The same cause has sapped their initiative. Too much depends on their supreme command. They have sacrificed the friendship of their late allies, Italy and Japan and Franco's Spain, for a new ally who, if we are defeated, will eat them up. Nazi Germany is doomed in any case. As for our own country, what I fear is not military defeat—I do not think that we will be defeated by Germany alone, and in any case a defeat in the field would not be the worse thing that could happen to us—but the loss of our national soul. If this country went Nazi for the sake of beating the German Nazis, victory in the field would be the worst fate. I am not confident about the immediate future of the thing we have called European civilization any more than yourself. But I am quite sure that in the long run the spirit in which you and I believe will create a better civilization, though you and I shall not see it . . .

Letter to T. M. Knox from R. G. Collingwood. Knox manuscripts University of St Andrews, MS 37524/430. Dated 6 Jan. 1940.

. . . at the beginning of term, finding that Paton had melted away and that there would be no professorial lectures on moral philosophy, I put on a course myself, and found it well and keenly attended. I suppose that in the near future numbers will diminish as men are called up; but I shall go on lecturing both in metaphysics and in moral philosophy as long as anybody comes, and think it far more important to do this than to obtain any alternative employment in the service of the government.

For it seems to me that we are engaged in a war of ideas, and that under the disadvantage of having lost the initiative. Nazi ideas have the explosive force of novelty: what we call 'democratic' ideas are old and stale, and—silent inter arma leges—likely to become absolutely decrepit in war conditions, whose effect might easily be the intellectual bankruptcy of our own side, even (perhaps I should say especially) in the event of military victory. People like you and me have a clear duty to prevent this if it can be prevented; and to diminish the evil effect of it if it can't.

Accordingly, as my audiences shrink, I mean to spend more and more of my time on a new book which I began a few weeks ago, in which the idea of a 'free' or 'civilized' society shall be expounded *ab initio* and developed dialectically by counter-attacking the Nazi attack on it. Greatly daring (for I doubt my power to bend the bow of Ulysses), I shall call it *The New Leviathan*! . . .

Index